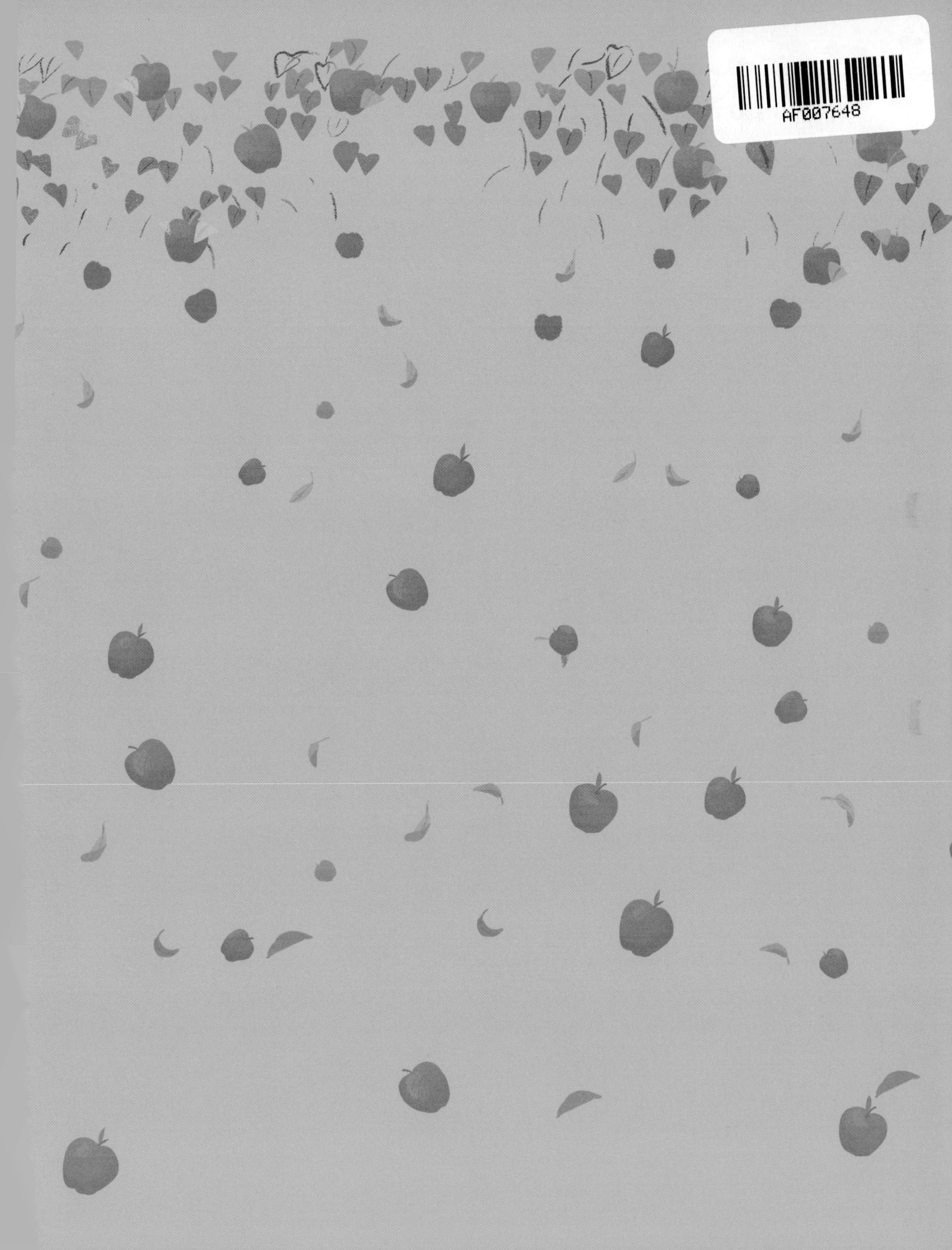

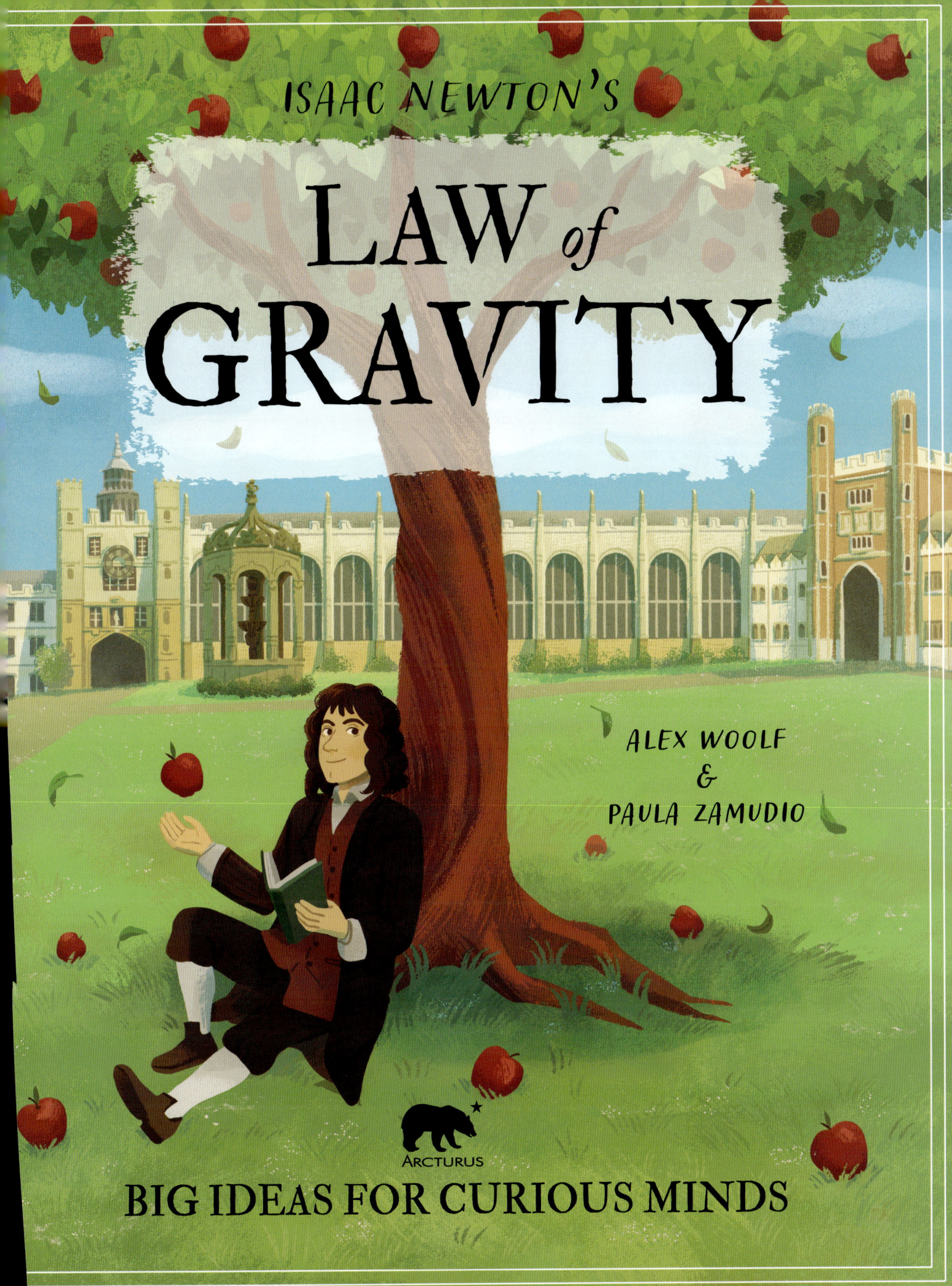

**ARCTURUS**

This edition published in 2024 by Arcturus Publishing Limited
26/27 Bickels Yard, 151–153 Bermondsey Street,
London SE1 3HA

Copyright © Arcturus Holdings Limited

All rights reserved. No part of this publication may be reproduced, stored in a retrieval system, or transmitted, in any form or by any means, electronic, mechanical, photocopying, recording, or otherwise, without written permission in accordance with the provisions of the Copyright Act 1956 (as amended). Any person or persons who do any unauthorized act in relation to this publication may be liable to criminal prosecution and civil claims for damages.

Writer: Alex Woolf
Illustrator: Paula Zamudio
Designer: Ariadne Ward
Consultant: Robert Snedden
Editors: Lydia Halliday and Lucy Doncaster
Managing Editor: Joe Harris
Managing Designer: Rosie Bellwood-Moyler

ISBN: 978-1-3988-3533-7
CH011594US
Supplier 29, Date 0824, PI 00007711

Printed in China

# ISAAC NEWTON'S Law of GRAVITY

## INTRODUCTION
What is Gravity?   4
Gravity Before Newton   6

## NEWTON'S LIFE
The Lonely Boy   8
From Farmer to Scholar   10
The Plague Years   12
Professor of Mathematics   14
International Fame   16
Final Years   18

## WHAT WAS NEWTON'S LAW OF GRAVITY?
The Moon and the Apple   20
The Cannon on the Mountain   22
Like Stones on an Invisible String   24
The Inverse Square Law   26
A Universal Force   28
The Three Laws of Motion   30

## NEWTON'S OTHER WORK
Calculus   32
Light and Color   34
The Reflecting Telescope   36
Alchemy   38

## NEWTON'S LEGACY
Newton versus Descartes   40
The Clockwork Universe   42
The Shape of Earth   44
The Tides   46
Space Exploration   48
Engineering   50
The Scientific Method   52

## THE STATE OF SCIENCE TODAY
Space and Time   54
Einstein's Theory of Gravity   56
Quantum Theory   58
What We Still Don't Know   60

**Glossary**   62

**Index**   64

# INTRODUCTION
## WHAT IS GRAVITY?

Gravity is an invisible force that pulls objects toward each other. It causes planets to orbit (travel along a circular or oval path) around the Sun, and causes the Moon to orbit around Earth. Gravity causes the tides, because the Moon pulls the seas toward it. Gravity is also the force that makes things fall to the ground when we drop them, and it is the force that keeps everything in place on Earth.

**Gravity and Weight**

When you stand on a weighing scale to check your weight, what you are actually measuring is the strength of the force of gravity acting on the mass, or physical stuff, in your body. The greater your mass, the greater your weight will be.

**How Powerful is Gravity?**

The more mass an object has, the more powerful its gravitational pull. An apple has a gravitational pull, but we can't feel it because an apple has very little mass. Earth's gravity is powerful because Earth has a very large mass. We can see this when we watch a rocket lift off—it needs an enormous amount of power to overcome Earth's gravitational pull and to reach space.

A huge amount of energy is needed for a rocket to overcome gravity and blast off into space.

## Grammar School

When Isaac was ten, Barnabas died and his mother returned to Woolsthorpe with her other children. Two years later, Isaac was sent to the grammar school (selective high school) in the nearby town of Grantham. There he learned Latin, and studied the Bible and classical literature.

He lived with William Clark, an apothecary (a seller of medicines). Clark taught Isaac how to mix up potions and remedies. Isaac neglected his studies at school and received low grades. In his spare time, he liked to build things, including a model windmill, a water clock, a four-wheeled cart, and paper lanterns.

IN HIS SPARE TIME, ISAAC ENJOYED BUILDING THINGS.

## Called Home

Following a fight in the schoolyard, Isaac's attitude to school changed and he began taking his studies more seriously. He even had hopes of going to university. However, when he was seventeen, his mother called him home to help manage her estate, which involved him becoming a farmer.

# FROM FARMER TO SCHOLAR

Isaac Newton turned out to be a terrible farmer. Instead of caring for his livestock or mending fences, he would build water wheels in the stream, or gather medicinal herbs while his sheep trampled over the barley. On market day in Grantham, he would leave a servant to run the stall so he could visit the library of his former landlord, William Clark.

### Trinity College, Cambridge

Newton's mother eventually realized farming wasn't for Isaac and let him complete his education. He returned to the grammar school at Grantham in 1660 and in June 1661 went on to Trinity College, Cambridge. The subjects at Cambridge were very traditional. Professors taught the theories of Aristotle and other ancient Greek philosophers as if they were fact. However, in the library, Newton found books that contained new ideas by scientists such as Copernicus, Kepler, Descartes, and Galileo, which challenged ancient theories. Copernicus, for example, argued that Earth traveled around the Sun, not the other way around, as the ancients had believed.

*"Plato [an ancient Greek philosopher] is my friend, Aristotle is my friend, but my best friend is truth."*
Isaac Newton, written in the margin of his notebook while he was a student at Trinity College, Cambridge.

The other factor affecting the strength of gravity is distance. The farther you are from an object, the weaker its gravitational pull. A climber on top of a high mountain will weigh a little less than they do at sea level, because they are farther from the center of Earth, so gravity is weaker.

## The Force of Gravity

Gravity is the most important force in the vast expanse of space. As well as keeping celestial bodies (objects in space) in orbit, it holds galaxies together and creates new stars and planets by pulling together the material from which they are made. Yet over much shorter distances, gravity is weaker than other forces, such as magnetism. You can see this if you place a nail on a table and hold a magnet just above it. The nail will defy gravity and attach itself to the magnet, because in this case, the force of magnetism is stronger.

## Newton and Gravity

The great scientist Isaac Newton was fascinated by this mysterious force. In 1666, he named it "gravity" from the Latin word *gravitas*, meaning "weight." Newton was the first person to understand that gravity is a universal force, affecting the motion of the Moon and planets, as well as objects here on Earth. He discovered a mathematical law to describe how gravity works.

# GRAVITY BEFORE NEWTON

Before the time of Isaac Newton, scientists had wondered for many centuries about the strange, invisible force that later came to be called gravity. What could possibly cause separate, unconnected objects to move toward or around each other? There is no string pulling the Moon around Earth, yet somehow it remains in its orbit.

## An Inner Heaviness

In the fourth century BCE, Greek thinker Aristotle theorized that objects had an inner heaviness that caused them to fall toward the center of the Universe (which he believed was Earth). But not all objects, he said, had this heaviness. He thought that fire and air, for example, moved upward.

Persian scholar al-Biruni (eleventh century CE) was among those who disagreed with Aristotle, and instead suggested that all heavenly bodies could attract objects, not just planet Earth.

ARISTOTLE

Aristotle believed that objects fell because each of the four elements (earth, air, fire, and water) tended to move toward their natural place. That's why air and fire are above us, and earth and water are below.

"Bodies fall towards Earth as it is the nature of Earth to attract bodies, just as it is in the nature of water to flow."
Indian astronomer Brahmagupta (c. 598–c. 668 CE)

## Galileo's Falling Experiment

Italian scientist Galileo Galilei (1564–1642) discovered, through experimenting, that falling objects accelerate at the same rate, regardless of their mass. In other words, two spheres of different masses dropped from a tower at the same time will also land at the same time. So, Galileo disproved Aristotle's theory that heavier objects fall faster than lighter ones.

GALILEO GALILEI

OLD IDEA

GALILEO'S DISCOVERY

## René Descartes and His Whirlpools of Ether

In 1644, French scientist and mathematician René Descartes (1596–1650) suggested that space isn't empty at all. Instead, he said it is filled with an invisible material known as "the ether" that moves in vast whirlpools, and that this is what causes the orbits of the planets (see pages 40–41).

This was the state of knowledge about gravity by the 1660s, when Isaac Newton first turned his mind to the subject.

# NEWTON'S LIFE
## THE LONELY BOY

Isaac Newton was born on Christmas Day 1642, three months after his father's death. He lived with his mother, Hannah Ayscough, at Woolsthorpe Manor in Lincolnshire, England. When Isaac was three, his mother married Reverend Barnabas Smith and went to live with him, leaving her son in the care of his grandmother, Margery.

### Dame School
Young Isaac must have felt lonely and abandoned by his mother, as she and Barnabas raised a family of three children in a nearby village. When he was old enough, Isaac went to the village dame School (a school for young children), where he learned to read and write.

ISAAC WAS A LONELY BOY WITH FEW FRIENDS.

### The Sundial
Isaac noticed how, on bright days, the sunlight crept steadily along the wall of his home at Woolsthorpe. He decided to mark its progress with wooden pegs hammered into the wall. As the days grew longer in spring, and shorter toward winter, he adjusted the position of the pegs. In this way, he created a sundial.

## Notebook of Questions

In his second year at Trinity, Newton began keeping a private notebook in which he asked himself questions about science. Newton came up with forty-five questions, and under each one he wrote down what was known about that subject. He then used reason to try to deduce (work out) new answers. His questions included: What is matter? Is space infinite? Why do objects fall?

## University Closes

In 1665, Newton sat his final examination and graduated with a Bachelor of Arts degree. He would have liked to continue with his studies, but that summer, the plague struck Cambridge and the university was forced to close.

# THE PLAGUE YEARS

Newton was twenty-two when he returned to Woolsthorpe, and he was a much more confident young man than the failed farmer who had left for Cambridge four years earlier. He was greeted warmly by his mother and three half-siblings, Benjamin, Mary, and Hannah. His grandmother, who had raised Isaac almost single-handedly, had sadly passed away.

## Powers of Concentration

Newton's mother gave him a 1,000-page notebook, and he used this to record his notes and experiments. The next two years were known as his "miracle years", because this was when he made some of his greatest discoveries.

He returned to university in March 1666 but plague struck again in June, so back he went to Woolsthorpe once more.

Thousands of people died in the plague of 1665–1666. Those who could, fled to the countryside.

NEWTON SPENT A LOT OF TIME THINKING IN THE ORCHARD AT WOOLSTHORPE.

## Major Discoveries

During the eighteen months he spent at Woolsthorpe, Newton made three major discoveries. First, he discovered a new form of mathematics (see below). Second, through his experiments with sunlight and prisms, he discovered that white light is made up of all the colors (see pages 34–35). And third, he discovered that gravity is a universal force that pulls objects toward each other (see pages 20–29). Its power can be worked out mathematically, based on the mass of the objects and their distance apart.

## A New Form of Mathematics

While Newton was at Cambridge, his mathematics tutor, Isaac Barrow, had told him that traditional mathematics described a motionless world. Now they needed a new kind of mathematics to describe change and motion. This would help make sense of things like the orbits of planets, or the motion of fluids. While at Woolsthorpe, Newton developed his "method of fluxions," later known as calculus (see pages 32–33).

"For in those days [the plague years of 1665–1666] I was in the prime of my age for invention & minded Mathematics & Philosophy more then than at any time since."
Isaac Newton

# PROFESSOR OF MATHEMATICS

Newton was very happy to return to Cambridge in 1667. He showed Isaac Barrow, his mathematics tutor, his method of fluxions. Barrow was astounded. Newton became a Fellow of Trinity College at the age of just twenty-four, and his career as a scholar had begun.

### Seeing the Light

In June 1669, Newton wrote a paper explaining his method of fluxions in more detail. As a result, Newton began to become better known. That same year, at the age of just twenty-seven, he became Professor of Mathematics. As a professor, Newton was required to give weekly lectures.

His lectures on light and color would later inspire his book *Opticks*, which was published in 1704. Through his study of optics (the science of sight and the way light behaves), Newton learned about lenses, which led him to invent a new type of telescope—the reflecting telescope (see pages 36–37).

NEWTON GIVING A LECTURE ON OPTICS.

## Newton Joins the Royal Society

Newton demonstrated his telescope to the Royal Society (RS), an organization of England's top scientists, in 1671. They were very impressed and, in 1672, Newton was elected a Fellow of the Royal Society.

NEWTON DEMONSTRATED HIS REFLECTING TELESCOPE TO OTHERS AT THE ROYAL SOCIETY.

## Rivalry with Hooke

The year 1672 also marked the beginning of a long and bitter rivalry between Newton and another member of the RS, Robert Hooke (1635–1703). Newton presented a paper in which he argued that light was made up of particles (see pages 34–35). Hooke disagreed, saying light was a wave. Discoveries in the twentieth century proved that both of them were right—light can behave as both a wave and a particle (see pages 58–59).

ROBERT HOOKE

# INTERNATIONAL FAME

In 1679, Robert Hooke wrote to Newton, telling him all about his theory on the motions of planets. This reminded Newton of his own discoveries at Woolsthorpe in 1666, and prompted him to continue with his work on gravity. The eventual result was a book that would win him worldwide fame.

### The *Principia*

In 1687, Newton published his very important book, *The Mathematical Principles of Natural Philosophy*. Known as the *Principia*, it presented a completely new system for understanding the Universe, introducing concepts such as gravity, force, and mass.

### Enemies for Life

The *Principia* made Newton famous all over the world, but it also reignited his rivalry with Robert Hooke, who accused Newton of stealing his theory about the motions of planets. It was true that, in his letter, Hooke had suggested a theory that Newton had used, but it was Newton who proved it mathematically. Newton and Hook continued to disagree on this, and they remained enemies until Hooke's death in 1703.

### New Directions

Following publication of the *Principia*, Newton wanted a new direction in life. He once again became involved in public affairs and was elected as Member of Parliament for Cambridge. He also began a study of alchemy, exploring how to change one kind of matter into another (see pages 38–39).

### Master of the Mint

In 1699, Newton was appointed Master of the Royal Mint, the place responsible for making English coins. At this time, there were many fake coins, or ones that had been interfered with in circulation. Newton called in all these coins and issued new ones. He dealt out harsh punishments to the counterfeiters (fake coin makers).

# FINAL YEARS

In 1703, Newton was elected president of the Royal Society. He would use this position to fiercely defend his discoveries and to hold power over other scientists. His final years were marked by conflict and disagreement.

In 1705, Newton was knighted by Queen Anne. He was the first scientist to receive such a recognition.

## The Calculus Clash

In the early 1700s, Newton clashed with the German mathematician Gottfried Leibniz (1646–1716) over the question of who first invented calculus. Each man accused the other of stealing his ideas. Today, most experts believe that they each invented calculus independently.

## Star Wars

In 1712, Newton got into a conflict with John Flamsteed (1646–1719), the Astronomer Royal. Flamsteed had collected a huge amount of information during a lifetime of stargazing. Newton wished to make use of Flamsteed's "Star Catalogue" for his own update of the *Principia*. Against Flamsteed's wishes, Newton published the catalogue.

## A Burial Fit for a King

Newton died in his sleep on 31 March 1727 at the age of 84. He was given a ceremonial funeral and was buried in Westminster Abbey, the traditional resting place of kings and queens. He was the first scientist to be buried there.

Newton's monument shows him resting on a pile of his books. The celestial globe (map of the stars in the sky) shows the constellations.

The famous poet Alexander Pope wrote this about Isaac Newton after his death:

"Nature and Nature's laws lay hid in night: God said, Let Newton be! and all was light."

# WHAT WAS NEWTON'S LAW OF GRAVITY?

## THE MOON AND THE APPLE

When he was an old man, Newton told the story of how he came to discover his law of gravity. It all began, he said, one day in 1666. He was sitting beneath an apple tree in the orchard at Woolsthorpe thinking about the Moon and wondering what force was causing it to orbit Earth, when he looked up and saw an apple fall from the tree.

"And the same year [1666] I began to think of gravity extending to ye orb of the Moon ..."
Isaac Newton

## A Flash of Insight

Newton thought about these two things—the Moon's orbit and the apple's fall. At first, they seemed very different, but perhaps they weren't. In a sudden flash of insight, he realized that the force that pulled the apple from the tree could be the same force that pulled the Moon around the Earth. They were both being pulled by gravity.

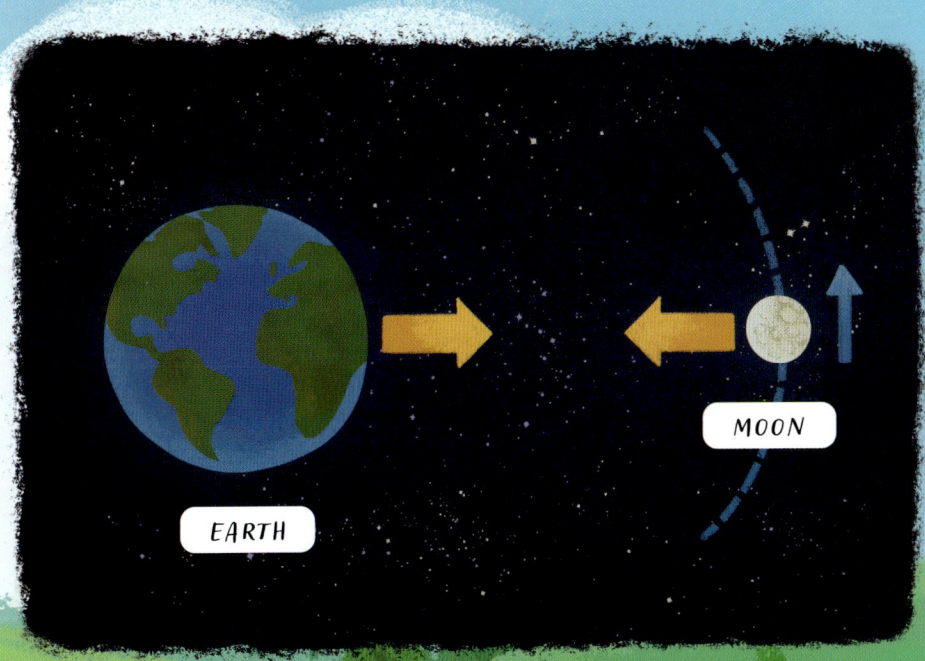

## A Universal Force?

Newton's rational mind didn't like this idea. The apple was close to the ground, while the Moon was very far away. How could the same force affect both objects? But then he thought, why shouldn't the effects of gravity extend out into space? Maybe gravity was a general force, acting on all the objects in the Universe, both large and small?

## Different Ways of Behaving

Then Newton thought of another problem. The apple falls straight to Earth, while the Moon circles it. How could the same force cause two objects to behave so differently? He started to imagine what might happen if he was to throw the apple high up into the sky. Would it also circle Earth, like the Moon?

# THE CANNON ON THE MOUNTAIN

Newton asked himself why the effects of gravity on the apple and on the Moon were so different. Why does gravity pull the apple straight down to Earth, yet cause the Moon to circle it? To explain how this could be, he came up with a thought experiment. He imagined placing a powerful cannon on top of a very high mountain.

## Firing Cannonballs

The peak of this imaginary mountain would have to be above Earth's atmosphere, so cannonballs fired from it would not be slowed by air resistance. Newton thought about what would happen if a cannonball was fired horizontally from the top of the mountain. He realized it would travel in a curved path before gravity eventually pulled it down to Earth. The faster the cannonball was fired, the farther it would travel before falling to the ground.

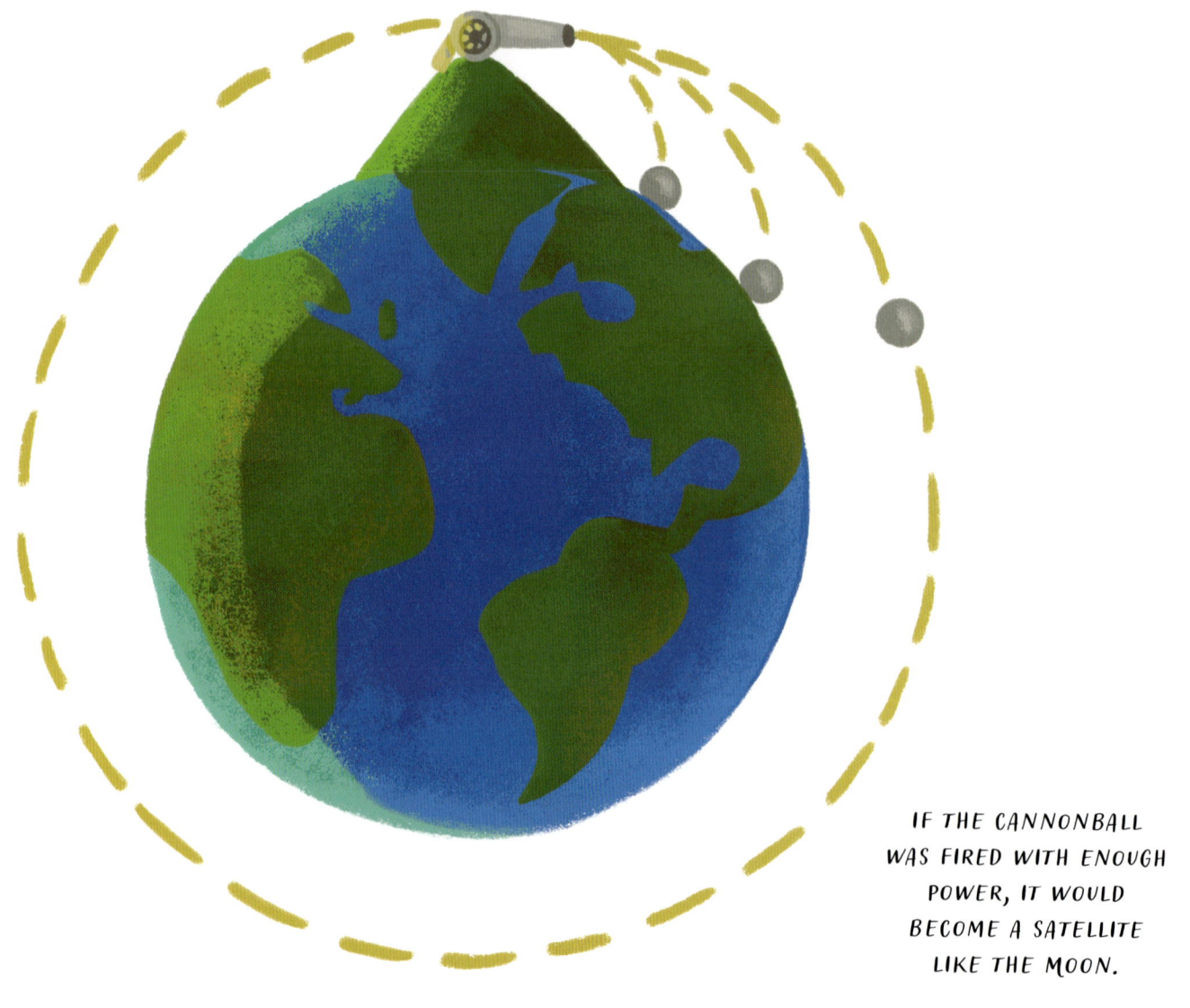

IF THE CANNONBALL WAS FIRED WITH ENOUGH POWER, IT WOULD BECOME A SATELLITE LIKE THE MOON.

## Falling Around Earth

But what if the cannonball was fired with so much power that it did not hit the ground but continued going around Earth? The cannonball would still be falling, but the rate at which it fell would be the same as the rate at which Earth's surface was curving away beneath it. So, the cannonball would be falling around Earth, not toward it, and it would become a satellite (an object that orbits a planet) like the Moon. So maybe the Moon is falling—not downward, but around the Earth.

*"In the celestial spaces above the Earth's atmosphere; in which spaces, where there is no air to resist their motions, all bodies will move with the greatest freedom."*
Isaac Newton, the *Principia*, 1687

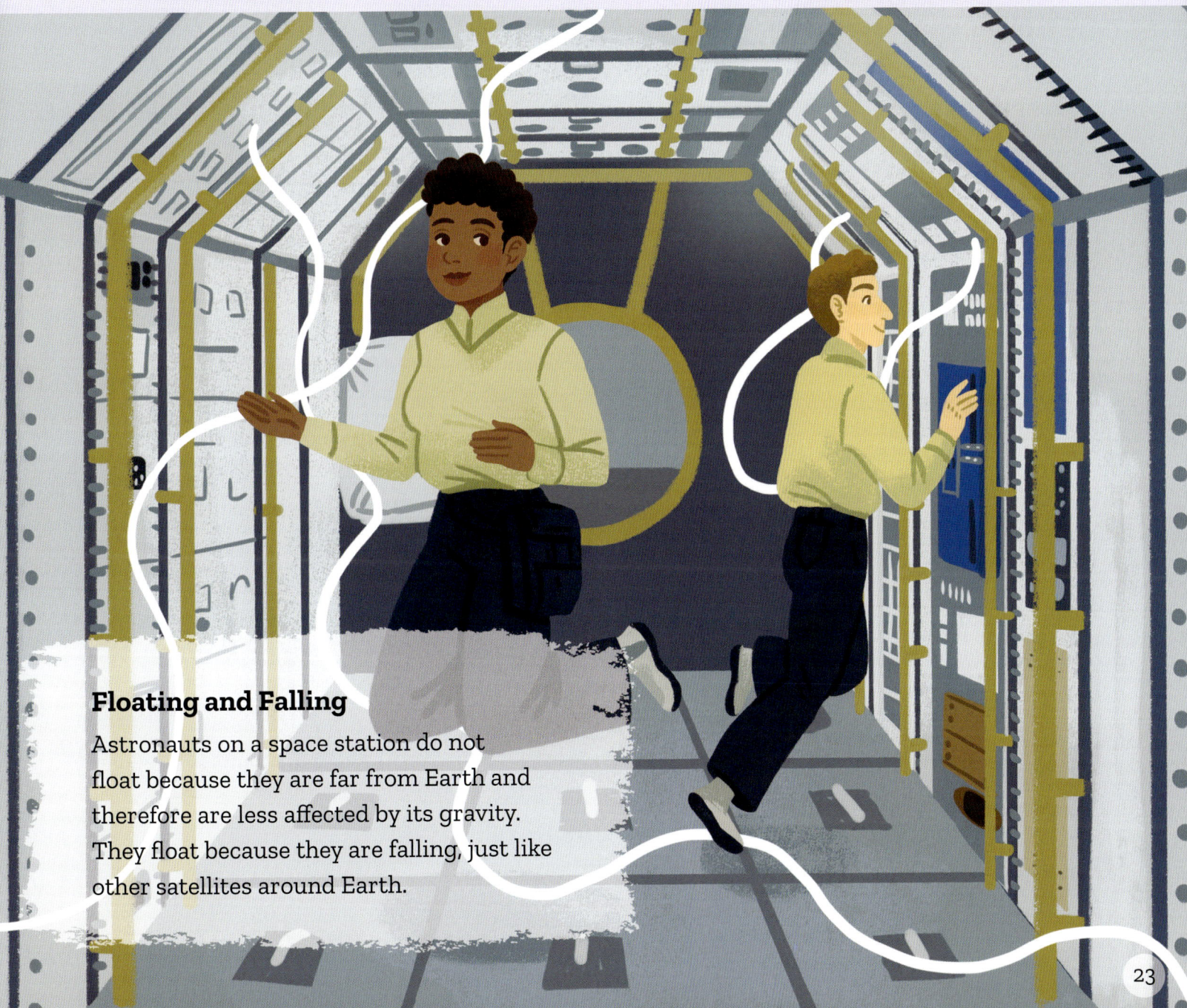

### Floating and Falling

Astronauts on a space station do not float because they are far from Earth and therefore are less affected by its gravity. They float because they are falling, just like other satellites around Earth.

# LIKE STONES ON AN INVISIBLE STRING

Newton worked hard on his theory of gravity between 1679 and 1684, eventually publishing it in the *Principia* (1687). He was helped in his work by the discoveries of previous scientists, including the Italian astronomer and physicist Galileo Galilei, and the German astronomer and mathematician Johannes Kepler (1571–1630).

## Learning from Others

From Galileo, Newton learned that an object in motion will keep moving in a straight line unless it is acted on by another force. In the case of the planets, Newton realized, this other force is the Sun's gravity, which changes the planets' straight-line movement into an orbit. From Kepler, Newton learned exactly how gravity acts on the planets. Kepler spent years studying information from studies of space, which helped him work out three laws of planetary motion.

JOHANNES KEPLER

### Kepler's First Law

The orbit of a planet is an ellipse (a kind of oval). This disproved the ancient Greek idea that heavenly bodies orbited in perfect circles.

### Kepler's Second Law

A planet does not go around the Sun at a constant speed. It speeds up when it is nearer the Sun, and slows down when it moves away from it. From this, Newton learned that gravity gets weaker with distance.

### Kepler's Third Law

The farther a planet is from the Sun, the slower it moves and the longer it takes to complete its orbit. This was further evidence that the power of gravity decreases with distance. What's more, Kepler showed that there was a precise mathematical relationship between the speed of a planet and its distance from the Sun. This proved to Newton that gravity follows a mathematical law.

SUN · MERCURY · VENUS · EARTH

### Centripetal Force

The gravitational pull of the Sun exerts a centripetal force on the planets. Centripetal means "center-seeking." We experience this force on Earth when we tie a stone to a piece of string and swing it around our head. The Sun's gravity is like an invisible string that holds onto the planets and stops them from flying off into space. If the string breaks, the stone will fly off in a straight line. In other words, the stone will do what all objects will do when not acted upon by another force.

IF THE STRING BREAKS, THE BALL WILL FLY OFF IN A STRAIGHT LINE.

# THE INVERSE SQUARE LAW

Galileo proved that all objects fall to Earth at the same rate, whatever their mass. Kepler's third law describing the motion of the planets appeared to show that gravity weakened with distance, and that it did so at a mathematically precise rate. Newton set about trying to find a way of resolving these two observations.

## Gravity Comes from the Center

On that day in the orchard in 1666, Newton noticed something very important about gravity—the apple fell straight down to the ground. It did not fall at an angle. This proves that gravity is exerted (applied) from the center of bodies, not from the sides. This became important later, when he did his calculations about gravity.

GRAVITY IS EXERTED FROM THE CENTER.

## A Mathematical Relationship

From Galileo, Newton learned that an apple falls from a tree at 9.8 m/s². Newton then worked out that the Moon falls toward Earth at 0.00272 m/s²—3,600 times slower. The Moon, he calculated, is 60 times farther from the center of the Earth than the apple is, and the force of gravity experienced by the Moon is 3,600 times weaker than that experienced by the apple.

This was exciting, because 3,600 is the square (a number multiplied by itself) of 60 (60 x 60 = 3,600). Newton knew he had found his mathematical relationship. As the distance between two objects increases, the gravity exerted by those objects decreases by the square of that distance. This is the "inverse square law." *Inverse* here means that if one thing (distance) increases, the other thing (gravity) decreases.

# How Gravity Lessens with Distance

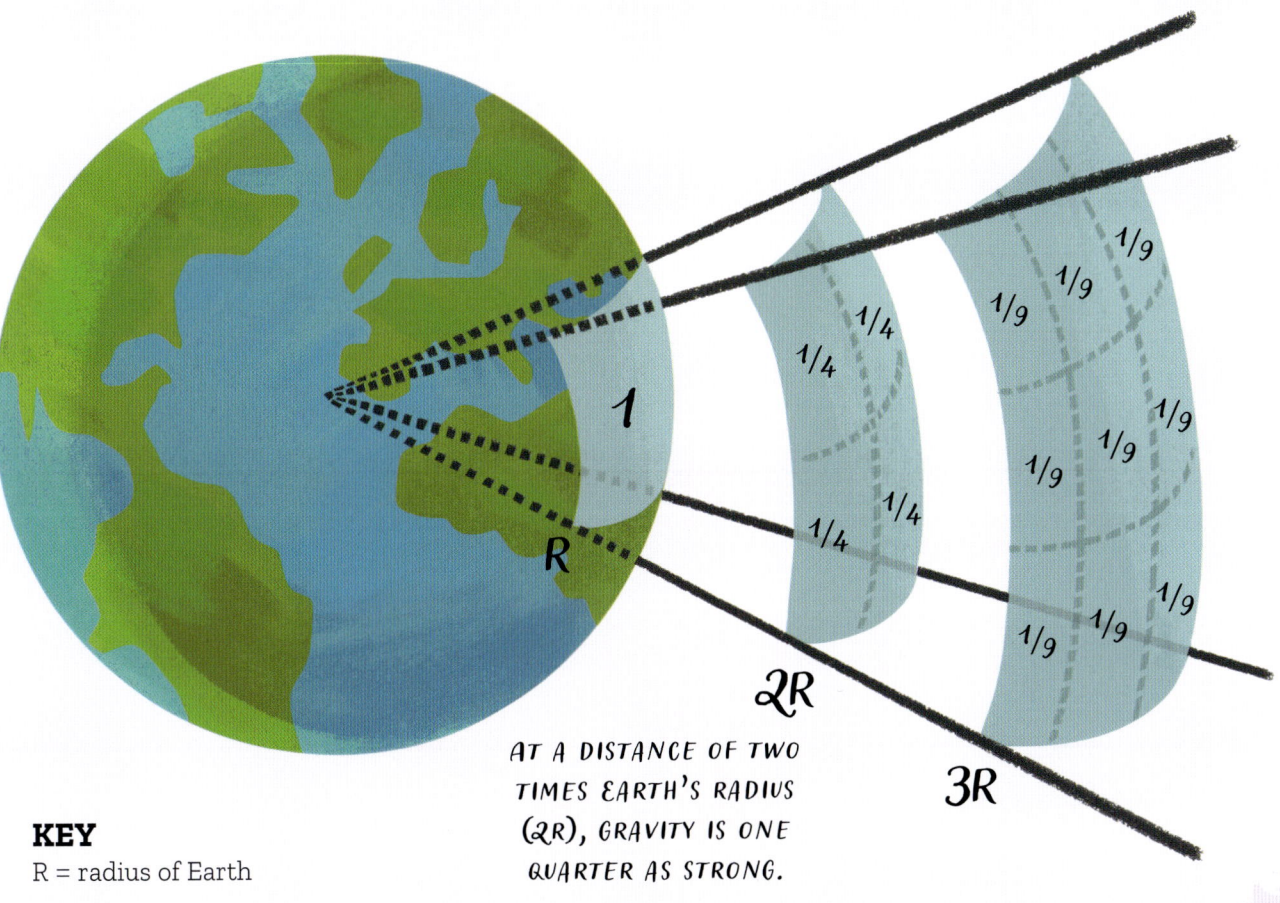

**KEY**
R = radius of Earth

AT A DISTANCE OF TWO TIMES EARTH'S RADIUS (2R), GRAVITY IS ONE QUARTER AS STRONG.

## Hooke's Contribution

Newton was not the first person to spot this relationship between gravity and distance. In his letters to Newton in 1679, Robert Hooke mentioned the inverse square law as a possibility. However, it was Newton who proved it mathematically.

"By such deductions the law of gravitation is rendered probable, that every particle attracts every other particle with a force which varies inversely as the square of the distance."
Isaac Newton, the *Principia*, 1687

# A UNIVERSAL FORCE

By linking the fall of an apple to the fall of the Moon in a single mathematical equation, Newton had shown that gravity is a universal force. This means gravity is at work between every lump of matter and every other lump of matter everywhere, from a grain of salt to the largest star.

Gravity works in both directions. You are pulled toward Earth, and Earth is also pulled toward you, but only by a very tiny amount!

**Gravity and Mass**

We have already looked at the relationship between distance and gravitational pull (see pages 26–27). Newton also calculated that an object's mass is directly related to the gravity it exerts. So if Object A has twice the mass of Object B, it will have twice the gravitational pull. To work out the gravitational pull between Object A and Object B you would need to multiply their masses, as well as measure the distance between them.

## A Weak Force

Newton came up with the Gravitational Constant—a measurement of the strength of gravity (G). G is an extremely tiny number. This is why the force of gravity is so weak at small scales (sizes), and why an apple isn't dragged toward your hand. It is only at the scale of stars and planets that gravity is powerful.

A planet's gravity turns it into a sphere, but only if it is bigger than 400–600 km (248–373 miles) across. That's why small asteroids aren't spherical.

## The Law of Universal Gravitation

The *Principia* contains Newton's law of universal gravitation. It states that every particle of matter in the Universe attracts every other particle of matter. The force of attraction depends on the particles' mass and the distance between them. The strength of the attractive force is proportional to the total of the particles' masses (so the greater the mass, the greater the force), and is inversely proportional to the square of the distance between their centers (so as the distance increases, the force decreases).

# THE THREE LAWS OF MOTION

Gravity explains how an apple falls to the ground, but there are other forces that can cause an apple or any other object to move, such as being thrown into the air. In the *Principia*, Newton wanted to define the rules that affect every kind of motion. Between them, these three laws describe the relationship between a physical object and the forces acting on it.

**The Law of Inertia**

The first law states that an object at rest remains at rest, and an object in motion remains in motion, moving at a constant speed and in a straight line unless a force acts on it. So, a ball will remain at rest unless it is kicked. It will then continue moving until a force, such as friction, slows it down.

A BALL AT REST.

A BALL BEING KICKED.

"For the basic problem of philosophy seems to be to discover the forces of nature from the phenomena of motions and then to demonstrate the other phenomena from these forces."
Isaac Newton, the Principia, 1687

A SMALL SUITCASE NEEDS LITTLE FORCE TO MAKE IT ACCELERATE.

A LARGER SUITCASE NEEDS MORE FORCE TO MAKE IT ACCELERATE.

## The Law of Acceleration

The second law states that the acceleration (a) of an object depends on the object's mass (m) and the amount of force (F) applied. Newton's equation for this is:

*force = mass × acceleration*
($F = ma$)

This can also be written as:

*acceleration = force ÷ mass*
($a = F/m$)

When you ride a bike, your acceleration depends on the force you apply by pushing on the pedals, divided by the combined mass of yourself and the bike.

## The Law of Action and Reaction

The third law states that whenever one object exerts a force on another object, the second object exerts an equal and opposite force on the first object. When you release an air-filled balloon, the air pushes out of the open end, causing the balloon to move forward. The force of the air escaping (the "action") is equal and opposite to the forward movement of the balloon (the "reaction").

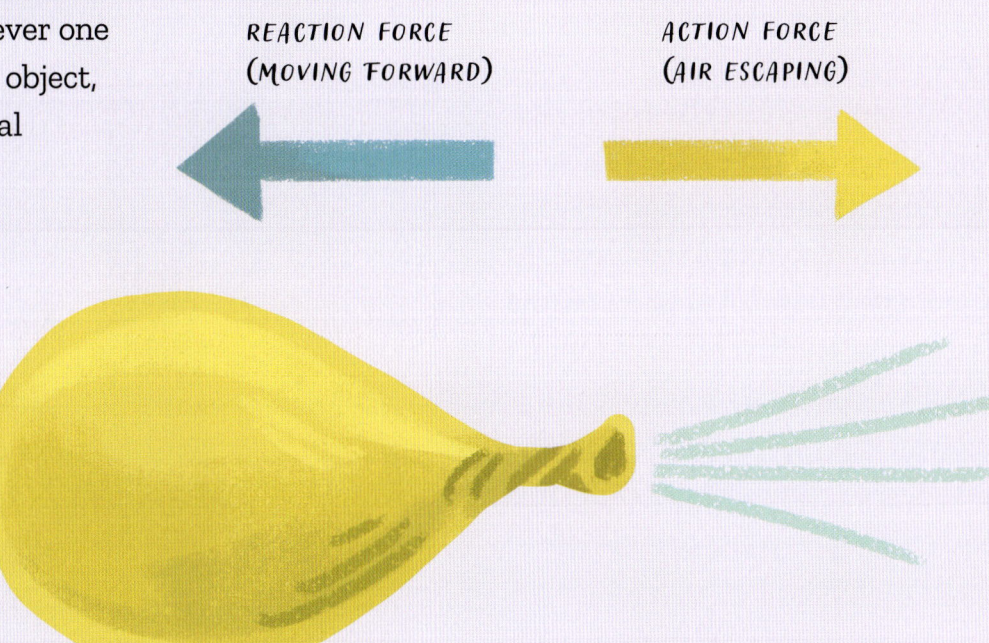

REACTION FORCE (MOVING FORWARD)

ACTION FORCE (AIR ESCAPING)

# NEWTON'S OTHER WORK
## CALCULUS

To help him with his work on gravity, Newton needed to develop a new kind of mathematics. He understood that if an apple falls, its speed will constantly increase until it hits the ground. In order to calculate the rate at which its speed is changing, he would need to measure the apple's speed an infinite number of times, which seemed impossible. He needed mathematics to describe the rate of change.

**Two Techniques**

If the apple fell at a constant rate, this could be shown as a straight line on a graph. But the apple's speed *increases* as it falls. This rate of change can be shown as a curve on a graph. A technique called differentiation can help work out the curve's slope.

Another technique, called integration, can be used to find the area (the amount of space inside a closed shape) beneath the curve. If you place a row of rectangles under the curve, you can calculate each of their areas. If you add those areas together, you can get an idea of the total area under the curve. The narrower the rectangles and the more of them there are, the more accurate the answer. But the rectangles would have to be infinitely (without a limit) narrow to give you a completely accurate figure.

AN APPLE'S FALL INVOLVES AN INFINITE NUMBER OF POINTS AT WHICH ITS SPEED IS INCREASING.

## Putting the Two Together

In 1665, Newton introduced a practical method of using differentiation and integration to calculate the rate at which something changes. He called it his "method of fluxions." It would come to be known as calculus.

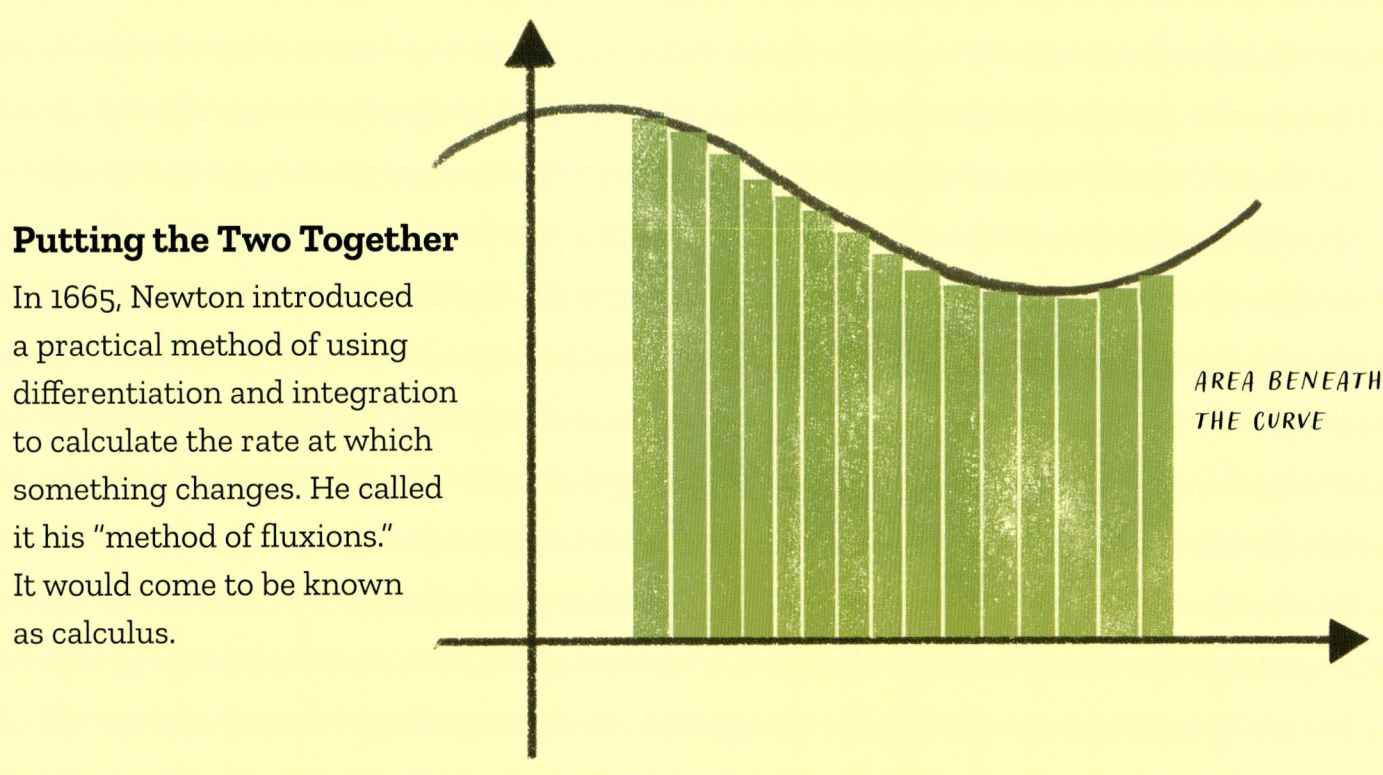

AREA BENEATH THE CURVE

## Infinite Series

Calculus requires you to make an infinite number of calculations in order to get a finite (limited in number or size) answer. This seems impossible, but in real life, doing something infinitely many times can lead to a finite result.

For example, when you cross a room, you first walk halfway across, then you walk halfway across the remaining distance, then half of that distance, and so on all the way to infinity. Yet despite this, if you keep going, you always reach the other side of the room. This idea of getting a finite answer from an infinite series is the central concept of calculus.

# LIGHT AND COLOR

Before Newton, there were many ideas about light and color. Some ancient Greek thinkers believed they were two separate things. Color was one of the properties of an object, and light—which had no color of its own—simply carried color from the object to the observer.

**Prisms and Rainbows**

Prisms are transparent (see-through) objects, often triangular in shape. When white light passes through a prism, it separates into a range of colors. Similarly, when sunlight reflects and refracts through raindrops, it causes a rainbow.

Natural prisms, in the form of rock crystals, have been known about since ancient times, and of course rainbows have been around forever. Until Newton, most people believed prisms and raindrops somehow added color to white light. They were wrong.

**Light-Bending Experiments**

At Woolsthorpe in 1666, Newton experimented with a prism. He passed a beam of sunlight through the prism and projected (shone) the colors onto a wall, taking careful note of their order and angle. Then he beamed these colors through a second prism so they turned back into white light. This proved that a prism doesn't add colors to white light as others had previously thought.

Newton was the first to understand that white light is actually made up of these colors. When light passes through a prism, it splits into the colors. Newton observed that light rays refract (bend) as they pass through the prism to form colors, and the color of each ray depends on the angle at which light leaves the prism. He realized that rainbows are caused by the same process—sunlight refracting through raindrops.

NEWTON EXPERIMENTING WITH A PRISM AT WOOLSTHORPE.

Through his experiments with prisms, Newton discovered that white light is made up of seven colors: red, orange, yellow, green, blue, indigo, and violet.

### Corpuscles

Newton believed that light is made up of tiny particles, which he named "corpuscles." In the twentieth century, physicists discovered he was right! They named the light particles "photons."

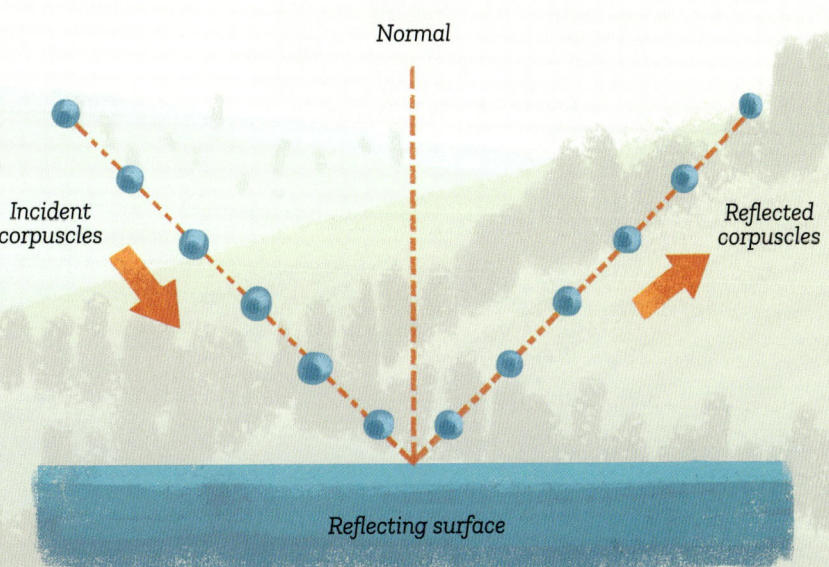

Newton believed that light was made up of particles rather than waves, because shadows form sharp edges, and because of the way light bounces off reflective surfaces.

# THE REFLECTING TELESCOPE

The telescopes of Newton's day were known as refracting telescopes. They had a lens at the front, then a long tube, then an eyepiece at the back. They worked by refracting light, so that parallel light rays met at a single point (the focal point), which made things appear larger.

Reflecting telescope

## The Problem with Lenses

Grinding glass to the right shape for telescope lenses was very difficult. Also, glass in Newton's day contained impurities that colored the lenses green. Worst of all, lenses behave a little like prisms, breaking white light into its individual colors, which caused colored halos around stars and other bright objects when they were looked at through a refracting telescope.

## Newton's Solution

Newton created a telescope that did not use lenses, but reflecting surfaces. He used two highly polished metal mirrors. The first mirror, at the bottom of the telescope tube, was concave (curved inward). Light from a star would strike this mirror and reflect back up the telescope tube, where it would strike a second mirror.

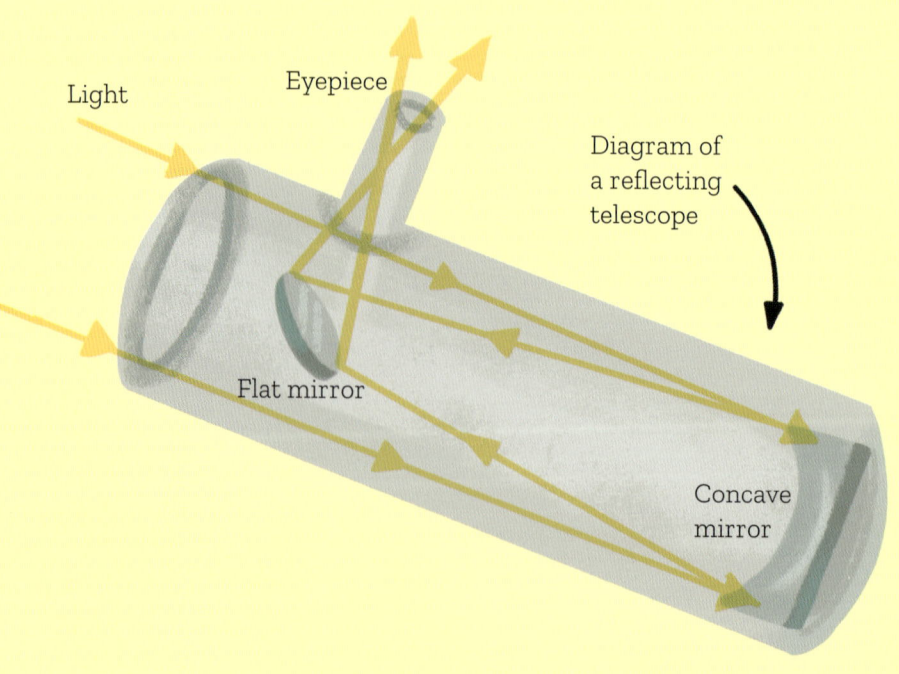

Diagram of a reflecting telescope

This second mirror, which was flat, was tilted at a 45-degree angle and reflected the light to an eyepiece at the side of the tube, where the viewer saw an image of the star.

## Small Yet Powerful

Newton's invention worked very well, not only getting rid of the halo effect, but also greatly increasing magnification (enlargement). His telescope magnified objects by around 40 times, giving it the same power as a much larger refracting telescope. It was successfully demonstrated to the Royal Society in 1671, then to King Charles II in 1672.

There was one problem with Newton's telescope, though. The highly curved first mirror caused a slightly distorted (misshapen) view. Later designs used mirrors with less of a curve, which solved the problem. By the eighteenth century, most astronomers were using reflecting telescopes.

The James Webb Space Telescope is a reflecting telescope built with a single segmented mirror to give it wide views of space.

# ALCHEMY

We mostly remember Isaac Newton for his achievements in physics, but he devoted far more of his time to alchemy. Alchemists were people who experimented with materials to try to turn base (non-valuable) metals into gold. This was an early form of chemistry—the study of substances and how they react to form new substances.

## A Desire to Understand

Newton's interest in alchemy was his attempt to understand the invisible forces that caused things to happen. For example, he wondered why heat is produced when two substances react, why when you comb your hair it causes the hair to rise up, or why water can rise up through paper, defying gravity.

In Newton's time, no one understood what electricity was, or knew about atoms or molecules. At first, he believed these effects were caused by an invisible substance called the ether. Later, he concluded that they were caused by attractions and repulsions between different particles of matter.

Newton understood the force of gravity that binds the Solar System, but not the forces that bind atoms. Yet he speculated about such forces in his writings.

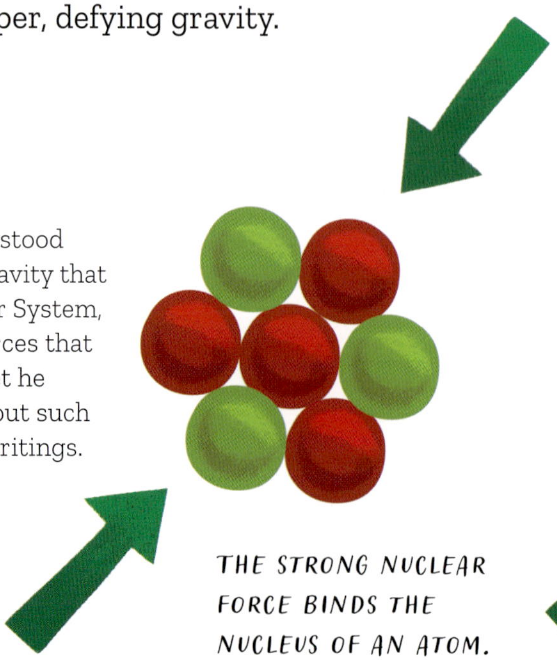

THE STRONG NUCLEAR FORCE BINDS THE NUCLEUS OF AN ATOM.

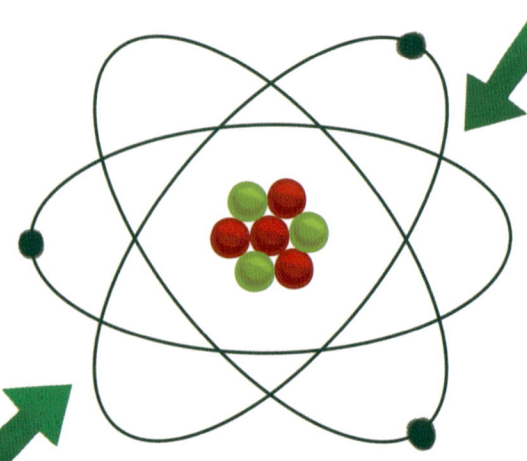

THE ELECTROMAGNETIC FORCE BINDS ATOMS.

## Experiments

Newton spent around thirty years studying alchemy. He did lots of experiments and wrote around a million words on the subject. Much of this is hard to understand because he used codes and obscure (unclear) symbols for the chemicals he experimented with. Yet modern scientists have tried to repeat some of the experiments and found that they work (though they don't produce gold!).

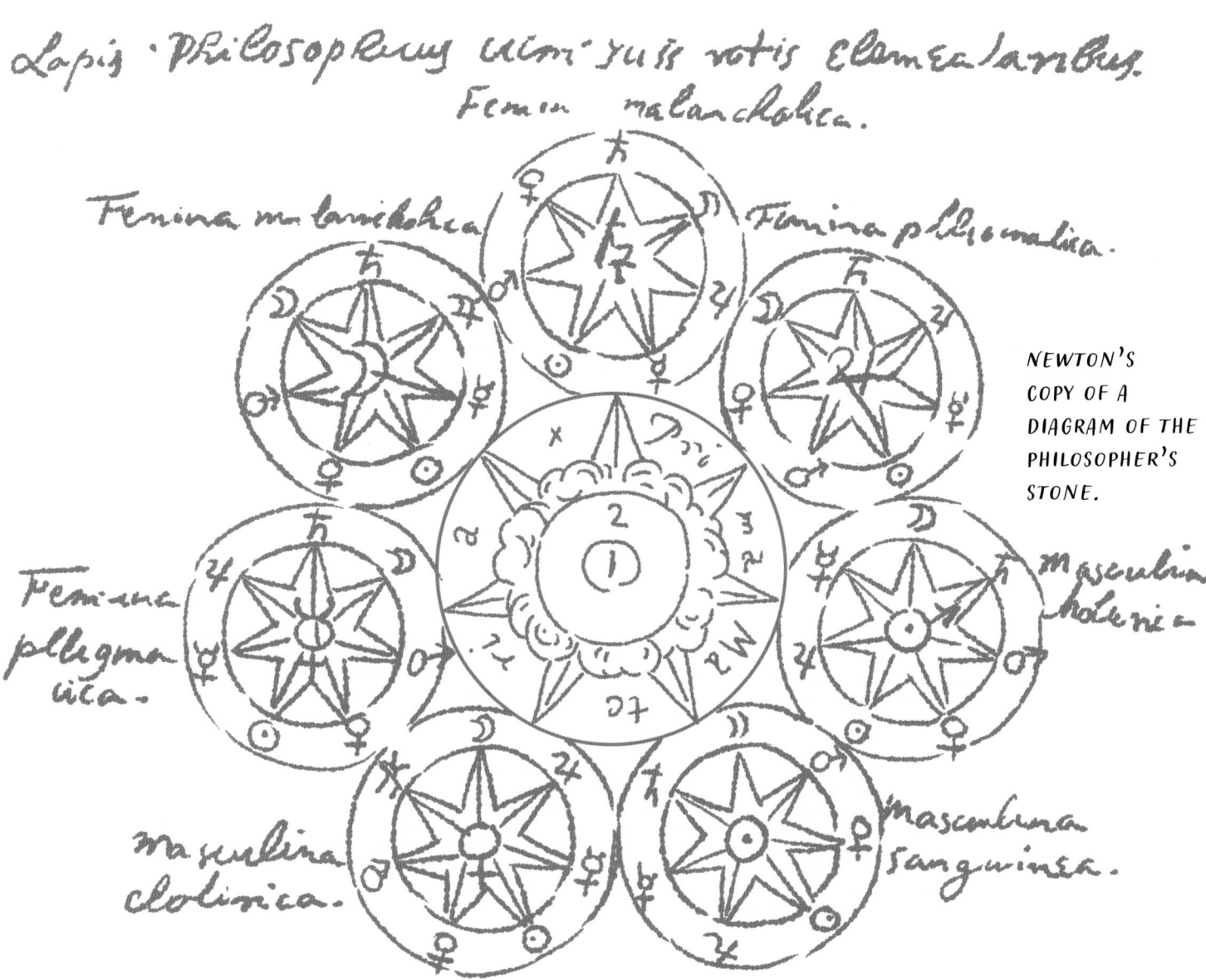

NEWTON'S COPY OF A DIAGRAM OF THE PHILOSOPHER'S STONE.

## A Valuable Outcome

The alchemists failed in their quest to turn base metals into gold, but their experiments did succeed in creating new alloys (mixtures of metals), acids, and pigments (dyes). They also invented equipment that would later be used for making perfumes and other products.

# NEWTON'S LEGACY

## NEWTON VERSUS DESCARTES

Isaac Newton was hailed as a genius in England during his own lifetime, but it took a little longer for his scientific discoveries to gain acceptance in Europe. This happened during the eighteenth century, thanks to enthusiastic supporters of Newton, including the French philosophers Émilie du Châtelet (1706–1749) and Voltaire (1694–1778).

VOLTAIRE

### Whirlpools of Ether

During the first part of the eighteenth century, most European thinkers still clung to the theory of gravity put forward by French scientist René Descartes. Descartes believed that space was not empty, because if it was, how could the Sun pull the planets around it? Instead, he suggested that the Universe was filled with an invisible substance called the ether. According to Descartes, the ether was constantly swirling around in huge whirlpools, and this was what caused planets to go around the Sun. To many Europeans, this seemed more likely than Newton's idea of gravity as an invisible force somehow acting on objects at a distance.

RENÉ DESCARTES

Descartes believed the Sun is at the center of a vast whirlpool of rotating invisible matter. The swirl of this whirlpool carries the planets around it, like leaves swirling as water empties into a drain.

## Forms of Reasoning

Descartes used a form of reasoning known as deduction. This involves using two general facts to reach a specific conclusion. For example, (1) the planets move around the Sun, and (2) nothing can affect something else at a distance. Conclusion—there must be some invisible substance between the planets and the Sun.

Newton used both deduction and induction to make his discoveries. With induction, you use specific facts to arrive at a general conclusion. His observation of a falling apple and the speed of the Moon led to a general conclusion—his law of gravity.

## Winning the Argument

In the end, Newton's view of gravity was accepted by the scientists of Europe because it explained things that Descartes' theory could not, such as the elliptical (oval) orbits of the planets, and Earth's tides.

*"If I have seen further it is by standing on the shoulders of giants."*
Isaac Newton, in a letter to Robert Hooke, 1675

# THE CLOCKWORK UNIVERSE

Newton saw the Universe as being governed by rational laws. His laws of motion and gravitation applied everywhere, from Earth to the most distant stars, suggesting a uniform, mechanical Universe. There was a comforting predictability and order to Newton's Universe, much like a vast clock.

## Classical Mechanics

The Newtonian model, known as classical mechanics, would dominate science for the next 200 years. According to classical mechanics, objects move under the influence of forces according to unchanging laws. So if the present state of an object is known, it is possible to calculate how it will move in the future and how it has moved in the past.

It was only in the late nineteenth and early twentieth centuries that the first cracks started to appear in classical mechanics. Its laws did not work at speeds approaching the speed of light, nor with objects at the tiny size of an atom (the smallest part of a chemical element). New theories were needed, such as Einstein's theories of relativity (see pages 54–57) and quantum theory (see pages 58–59) in order to provide a more complete understanding of the Universe.

Invented by French physicist Edme Mariotte (1620–1684) in the seventeenth century, Newton's cradle demonstrates Newton's first law of motion. The balls are at rest until they are struck by the swinging ball.

## The Enlightenment

Newton's vision of a Universe ruled by rational laws came to be known as Newtonianism. It helped inspire a movement called the Enlightenment, which lasted from about 1680 to 1820. Dates vary, but the publication of Newton's *Principia* (1687) is often regarded as the starting point of this movement. Enlightenment thinkers proposed that knowledge should be obtained through the use of reason and the evidence of the senses.

KING LOUIS XVI AT THE GUILLOTINE.

## Political Change

Many Enlightenment thinkers wished to introduce Newtonian ideas into politics. Newton's clockwork Universe emphasized harmony, balance, stability, and peace. This contrasted with the politics of the time, with wars raging across Europe and with all the power in the hands of kings and nobles. Many yearned for an ideal society based on Newtonian principles, in which all people could dwell under a single set of laws that guaranteed their freedom, equality, and natural rights.

One outcome of the Enlightenment was the violent overthrowing of the monarchy in France during the revolution of 1789. The king and thousands of nobles had their heads chopped off at the guillotine.

# THE SHAPE OF EARTH

Newton is not only remembered for discovering the laws of gravity and motion, he also calculated the shape of our planet. In the late seventeenth and early eighteenth centuries, the Royal Society in London and the French Academy of Sciences in Paris became locked in a battle over Earth's shape. Both agreed that it wasn't quite spherical, but they disagreed on exactly what shape it was.

### Prolate or Oblate?

The French, led by the mathematician and astronomer Dominique Cassini (1625–1712) and his son Jacques (1677–1756), argued that Earth was prolate (longer at the poles), like a lemon. However, the British, led by Isaac Newton, argued that it was oblate (wider at the equator and flattened at the poles), like an orange.

OBLATE

PROLATE

## Richer's Discovery

Newton worked out that the Earth is oblate thanks to information from French scientist Jean Richer (1630–1696). In 1671, Richer went to Cayenne in French Guiana, near the equator, and discovered that his pendulum clock lost an average of 2.5 minutes per day compared to Paris time (farther away from the equator). This made Newton think that the centrifugal effect (a force pushing outward from the center) caused by Earth's spin was strongest at the equator. This meant gravity must decrease as you move toward the equator, causing the planet to bulge outward at its middle.

## Two Expeditions

To resolve the issue, in 1735, the French Academy launched two expeditions to measure Earth's curvature at the poles and the equator. The polar expedition went to Lapland and the equatorial one went to Peru. They reported back a little over a year later with their results. Gravity, it turned out, is about 0.5 percent stronger at the poles than it is at the equator. So Newton was right—Earth has an oblate shape. On hearing the result, Voltaire congratulated the expedition leader "for flattening the Earth and the Cassinis."

## Consequences

Due to Earth's bulging middle, the surface of Earth at the equator is roughly 12 km (7.5 miles) farther from its center than it is at the poles. Also, because gravity is a tiny bit stronger at the poles than it is at the equator, a person weighing 100 kg (220 lb) in northern Canada can lose 500 g (1.1 lb) by moving to the Congo. Their mass stays the same, but their weight reduces because the force of gravity is less.

# THE TIDES

About twice every 24 hours, the oceans rise and fall. We call these the tides. In 330 BCE, ancient Greek explorer Pytheas (359–289 BCE) suggested that the Moon might cause the tides but could not say how. We can thank Newton and his law of gravity for solving the mystery. He showed how the tides are caused by the gravitational pull of the Moon and, to a lesser extent, the Sun.

PYTHEAS

## The Moon and the Tides

Newton understood that in the part of the ocean that is closest to the Moon, the ocean bulges upward. When the bulge reaches a coast, the water level rises. We call this high tide.

On the opposite side of Earth, a similar bulging of the ocean takes place, caused by Earth's circular movement. These two bulges explain why there are two high tides about every 24 hours.

HOW THE MOON AFFECTS EARTH'S TIDES

The Moon's gravity affects the watery parts of Earth because water, unlike rock, is free to move.

## The Sun and the Tides

The Sun is 27 million times more massive than the Moon, but it is also 390 times farther away from Earth than the Moon is. The result of this is that the Sun's gravitational pull on Earth, and its effect on Earth's tides, is just under half (46 percent) that of the Moon's. Solar (Sun-generated) tides are considered variations on lunar (Moon-generated) tides.

## Spring and Neap Tides

Just after a full and new moon, the Sun and Moon are lined up, and their combined gravity pulls the ocean into an even bigger bulge. We call this a spring tide. Seven days later, we have a neap tide, when the tides are at their lowest. This is when the Sun and Moon are at right angles to each other, cancelling each other out.

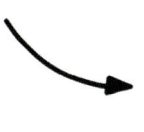

### SPRING TIDE

Spring tide, when the Sun and Moon are aligned. Their gravitational pull is combined, so tides are at their highest.

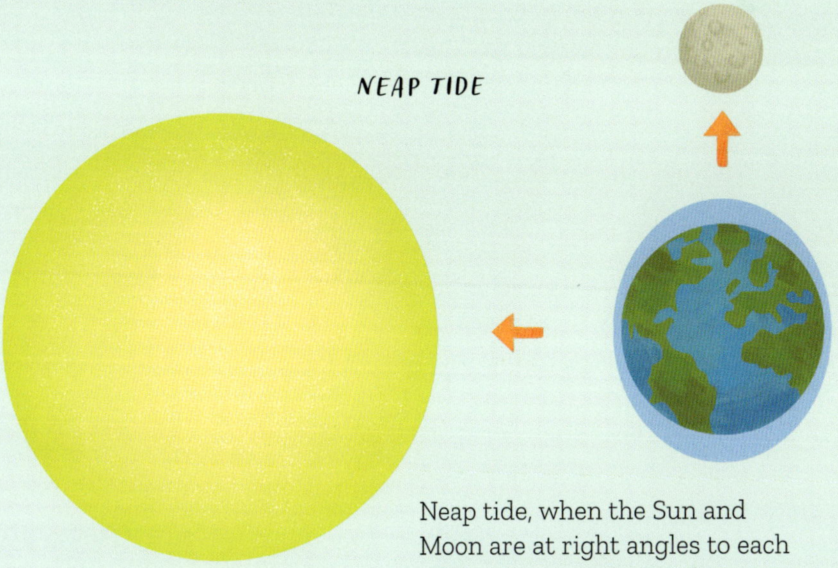

### NEAP TIDE

Neap tide, when the Sun and Moon are at right angles to each other. This cancels out their gravitational pull, so tides are at their lowest.

# SPACE EXPLORATION

Isaac Newton has had a major impact on space exploration, both in terms of astronomy and spaceflight. His laws of motion and gravitation together describe the motions of celestial bodies. Also, Newton's reflecting telescope design forms the basis of modern telescopes that allow us to see far into space.

### Predicting a Solar Eclipse

Even during Newton's lifetime, his law of universal gravitation was being used to predict astronomical events. His friend, the astronomer Edmond Halley, used Newton's law to make the first accurate prediction of a solar eclipse (when the Sun is blocked out by the Moon). Halley calculated that the eclipse would be visible across a large area of England on 3 May 1715, reaching London at 9.13 a.m. He was accurate to within four minutes.

### Halley's Comet

In 1682, a bright comet appeared in the sky. Edmond Halley identified it as the same one that had been seen in 1607 and 1531. The comet, he said, reappeared every 74–79 years. Using Newton's law of gravity, Halley predicted the comet's return in 1758. He was correct, and it was named Halley's Comet for him.

EDMOND HALLEY

"The motions of the Comets are exceeding regular, are govern'd by the same laws with the motions of the Planets ..."
Isaac Newton, the Principia, 1687

## A New Planet

Newton's law of gravity helped discover the planet Neptune. In 1821, French astronomer Alexis Bouvard (1767–1843) was studying the planet Uranus when he noticed a difference between his predictions of the planet's orbit, based on Newton's law, and what he was seeing through his telescope. He believed the difference must be due to the gravitational pull of a new planet. The mystery planet was first observed by German astronomer Johann Galle (1812–1910) in 1846, and was named Neptune.

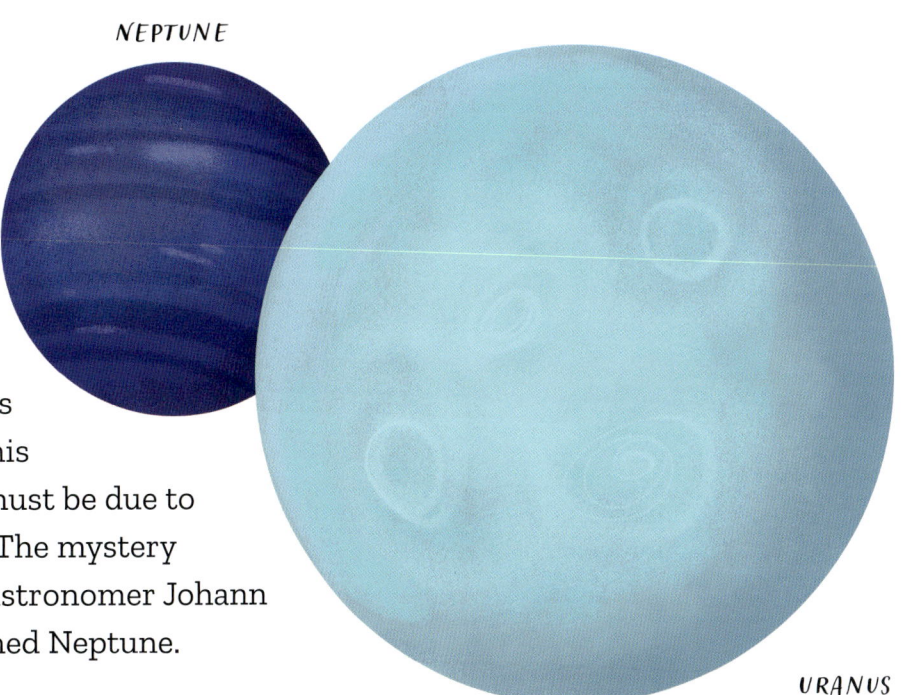

NEPTUNE

URANUS

## Apollo Missions

Newton's laws of motion were used during the Apollo missions to the Moon. For example, the second law (force = mass × acceleration) was used to calculate the forces required to accelerate the spacecraft during launch and to move about once in space.

Craters on the Moon and Mars have been named after Newton.

Apollo 11 command and service module.

# ENGINEERING

Isaac Newton's laws of motion and gravitation define, in mathematical terms, how forces act on objects. Engineers must take these laws into account when designing things like buildings, bridges, cars, or airplanes, to ensure their designs are safe and practical to use, and can cope with the forces they will experience.

**Air Bags**

Newton's first law of motion states that an object remains at rest or moves in a straight line at a constant speed unless acted upon by an external force. Engineers make use of this law when designing air bags in cars. If a car crashes, the driver will continue moving forward until an external force—the air bag—acts on them. Air bags (and also seat belts) increase the time it takes for the driver to decelerate (slow down) to zero. The longer the deceleration takes, the smaller the force on the driver's body, decreasing the chance of serious injury.

Newton's third law states that for every action there is an equal and opposite reaction. With a suspension bridge, the force of gravity pulling it down is equal to the pulling forces in the cables tugging it up, keeping the bridge stable.

## Calculating the Stresses

Newton's third law of motion states that when one object exerts a force on another object, the second object exerts an equal and opposite force on the first object. Engineers use this law when calculating the stresses acting on cables connecting train carriages or joints in the framework of buildings and bridges.

## Racing Cars and Trains

Newton's second law of motion states that an object's acceleration depends on its mass and the amount of force applied. This is why racing car designers are always looking for ways to reduce a vehicle's mass, so they can increase its acceleration. Engineers also apply the second law in train design, where Newton's equation (force = mass × acceleration) tells them the force they will need to pull a set of carriages of a certain mass.

## Flying Vehicles

Engineers use Newton's laws of gravity and motion when calculating the amount of lifting force required to overcome Earth's gravitational pull. This might be used for a rocket, an airplane, or a helicopter, for example. Similarly, they must take these laws into account when computing the trajectory (path) that a missile requires to reach a particular target.

# THE SCIENTIFIC METHOD

As well as making many great discoveries, Newton invented a scientific method that is still in use today. Long before Newton, people had carried out scientific experiments, including the thirteenth-century English philospher Roger Bacon (1220–1292). However, it was Newton who first included experimentation as part of a systematic (orderly) method for reaching scientific conclusions.

## From Question to Theory

The method that Newton followed, for example with his work on light and color, involved five essential stages. First, ask a question. Second, do some research. Third, come up with a theory. Fourth, carry out experiments to test the theory, carefully noting the results. And fifth, describe the experiment so that other scientists can replicate (copy) every step of the process. This method, involving a balance between theory and experiment, has served as a model for scientific investigations ever since.

QUESTION

RESEARCH

HYPOTHESIS

EXPERIMENT

CONCLUSION

## The Speed of Sound

A good example of Newton's scientific method in action is his measurement of the speed of sound in 1686. Not only was he the first person to measure it, but he also recorded exactly how it was done. It involved a theory (a formula for obtaining the speed of sound) and an experiment, which he performed in a covered walkway at Trinity College, Cambridge. He clapped his hands and measured how long it took for the sound to echo back to his ears. His figure was off by around 15 percent, due to inaccurate timing (he used a pendulum) and a slight error in his formula.

*"This Analysis consists in making Experiments and Observations, and in drawing general Conclusions from them by Induction."*
Isaac Newton, Opticks, 1704

## Rules of Reasoning

In his book *Principia*, Newton set down four rules of scientific reasoning:

1. The simplest possible cause of a phenomenon (something observed) is usually the correct one.

2. The same phenomenon seen in different places usually has the same cause (e.g. the orbits of planets and moons have the same cause—gravity).

3. Theories proven to apply everywhere on Earth must apply to the whole Universe (e.g. the laws of motion).

4. A conclusion reached by experiment is accurate until it has been disproven by another experiment.

# THE STATE OF SCIENCE TODAY

## SPACE AND TIME

Isaac Newton transformed science with his discoveries, but he was not right about everything. In this section we will look at some of the things Newton got wrong, based on our understanding of the Universe today.

### Absolute and Unvarying

Newton believed in absolute space and time. In other words, he believed that space would exist even if there were no objects in the Universe, and time would exist even if there were no events. In Newton's view, space and time are constant wherever you are in the cosmos. Time moves at the same speed, and space is always the same, and is not affected by the objects within it.

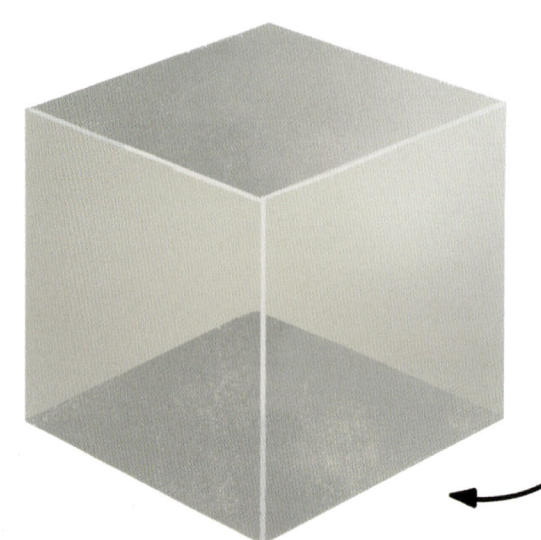

For Newton, space was like an invisible box, inside which objects existed. The objects didn't affect the box and weren't affected by it. Newton thought that if you took away the objects, the box would still be there.

> "Absolute, true, and mathematical time, in and of itself and of its own nature, without reference to anything external, flows uniformly."
> Isaac Newton, the *Principia*, 1687

## It's All Relative

In 1905, German scientist Albert Einstein (1879–1955) proposed a new theory that overturned Newton's idea about absolute space and time. With his special theory of relativity, Einstein showed that the speed of light—300,000 km (186,411 miles) per second in a vacuum—is always the same no matter how you measure it. So, whether you are motionless or moving at 299,000 km/second (185,790 mi/second), a light beam will always pass you at the same speed. For this to make sense, Einstein realized that time and space must vary depending on our speed. As someone approaches the speed of light, time must slow down and objects must shrink, when measured by someone who is not moving.

A motionless observer watching someone flying at close to the speed of light would observe the clock on board the flier's rocket ship moving more slowly. The rocket ship would also appear to shrink in the direction of travel.

### Einstein was Correct

Einstein's special theory of relativity has been shown by repeated experiments to be correct, proving that space and time vary for people depending on how fast they are moving. In other words, space and time are not absolute, as Newton believed, but *relative*. We don't notice this because the slowing of time and the shrinking of objects is extremely small at normal speeds. They would only become noticeable if we approached the speed of light.

# EINSTEIN'S THEORY OF GRAVITY

Newton believed that gravity is a force of attraction that exists between all objects with mass. His law of universal gravitation explains everything from the orbit of the Moon to the fall of an apple. However, there were problems with it. The orbit of the planet Mercury, for example, does not quite follow Newton's law. This mystery was eventually solved by Albert Einstein when he suggested a different model of gravity.

ALBERT EINSTEIN

## Warping Spacetime

With his general theory of relativity, published in 1915, Einstein proposed that gravity was not a force of attraction but a result of the warping of space and time by objects with mass. Space and time, he said, aren't separate, but are in fact one thing, called spacetime. Gravity is simply the effect of this warping of spacetime.

## Mercury's Odd Orbit

The perihelion of Mercury (the point in its orbit where it's closest to the Sun) moves forward about twice as far as Newton predicted each time it circles the Sun. Einstein showed that during part of Mercury's orbit it falls into the Sun's "gravity well"—a severe dip in spacetime—which is why the perihelion advances so far each time.

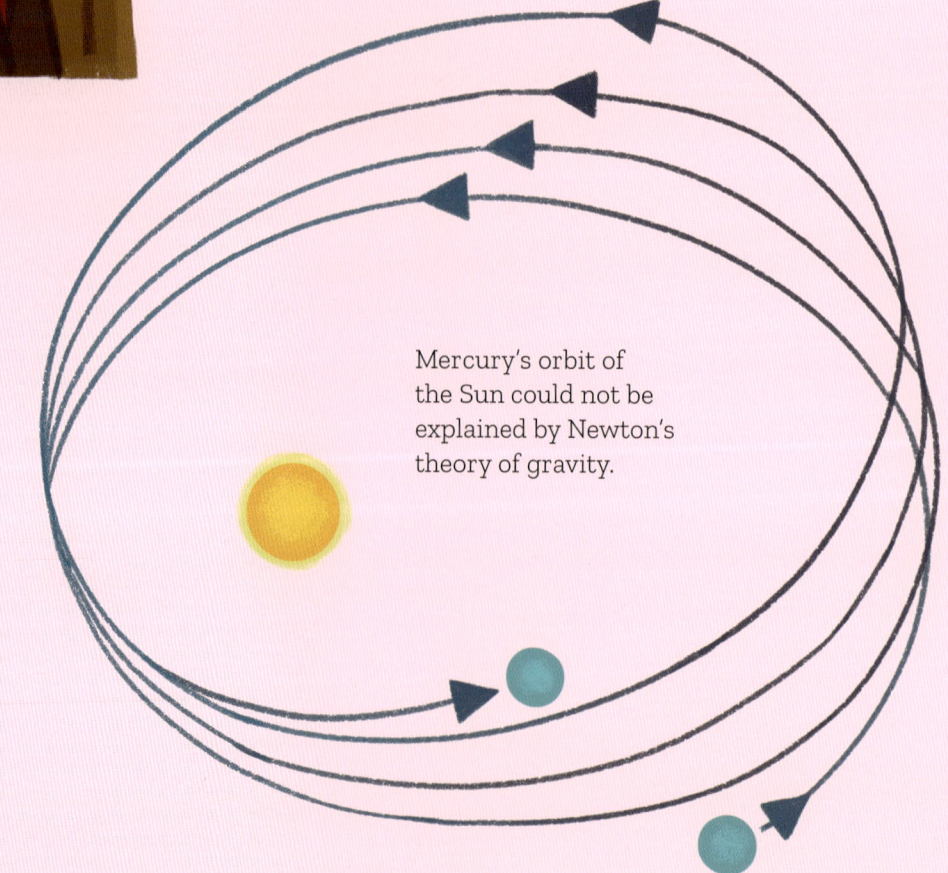

Mercury's orbit of the Sun could not be explained by Newton's theory of gravity.

## Photographic Evidence

Newton believed that space and time were absolute and were not affected by the objects that moved through them. With his general theory of relativity, Einstein showed that this was wrong. Objects *do* affect spacetime because their mass warps it. This was proved by British astronomer Arthur Eddington (1882–1944) in 1919 when he photographed stars during a solar eclipse. Later examination showed that the light from these stars had been deflected (made to change direction) due to the Sun's warping of spacetime.

Einstein asked us to imagine spacetime as a stretchy rubber sheet. Objects with mass rest on this sheet, creating dips in its surface. This warping explains why the Moon circles the Earth.

# QUANTUM THEORY

Newton believed in a Universe of cause and effect, where forces caused objects to move in predictable ways. However, in the early twentieth century, scientists discovered that at the very small scale of atoms, known as the quantum scale, everything becomes very unpredictable. Newtonian physics, or classical mechanics, breaks down at this level, and a quite different set of laws prevails.

### The Nature of Light

In 1801, British physicist Thomas Young (1773–1829) performed what became known as the double-slit experiment, which showed light behaving as a wave. This appeared to contradict Newton's theory that light is made up of particles. However, in 1905, Einstein showed that light is, in fact, made up of particles, later called photons. It turns out that light can be both a wave and a particle!

In the double-slit experiment, a light beam was shone through two parallel slits onto a screen. The pattern on the screen had fringes similar to water ripples, which seemed to prove that light was made up of waves.

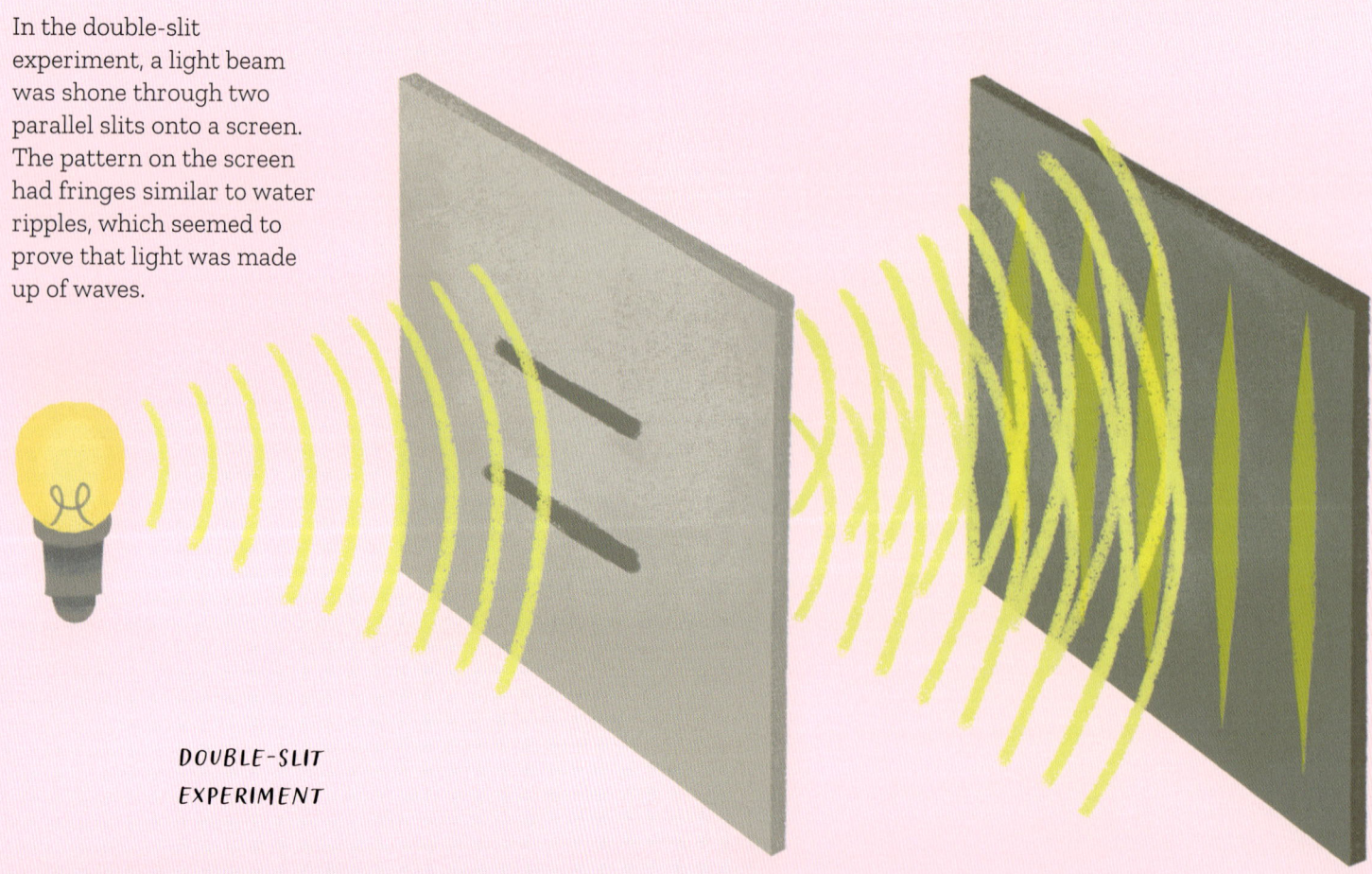

DOUBLE-SLIT EXPERIMENT

## Matter Waves

In the 1920s, physicists discovered that at the quantum level, matter can also behave like a wave. Electrons are negative particles that surround the nucleus (core) of an atom. Strangely, they only occupy a location in space when they are being observed. The rest of the time, their position is fuzzy and uncertain, like a wave.

"All material Things seem to have been composed of the hard and solid Particles."
Isaac Newton, Opticks, 1704

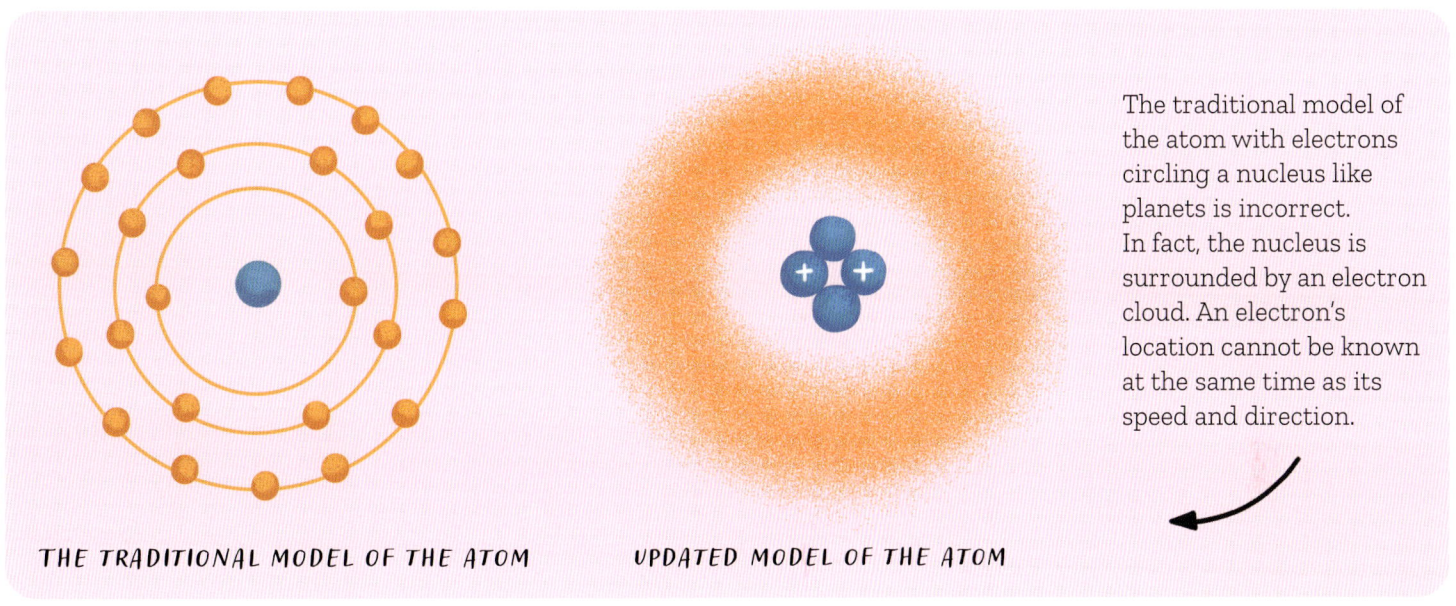

THE TRADITIONAL MODEL OF THE ATOM

UPDATED MODEL OF THE ATOM

The traditional model of the atom with electrons circling a nucleus like planets is incorrect. In fact, the nucleus is surrounded by an electron cloud. An electron's location cannot be known at the same time as its speed and direction.

## Schrödinger's Equation

In 1926, Austrian physicist Erwin Schrödinger (1887–1961) created an equation that describes mathematically the wave-like nature of electrons. The Schrödinger equation has the same importance to quantum mechanics as Newton's laws of gravity and motion have for classical mechanics. Newton's laws have not been replaced by quantum theory—they still work very well on the larger scales of apples and moons.

ERWIN SCHRÖDINGER

LOUIS DE BROGLIE

WERNER HEISENBERG

Werner Heisenberg (1901–1976) and Louis de Broglie (1892–1987) helped develop quantum theory in the 1920s.

# WHAT WE STILL DON'T KNOW

Newton discovered the law of universal gravitation, but he admitted he did not know what gravity was or what caused it. Einstein explained that gravity is simply geometry—the result of a warping of spacetime by objects with mass. Yet there's still plenty we don't know about gravity.

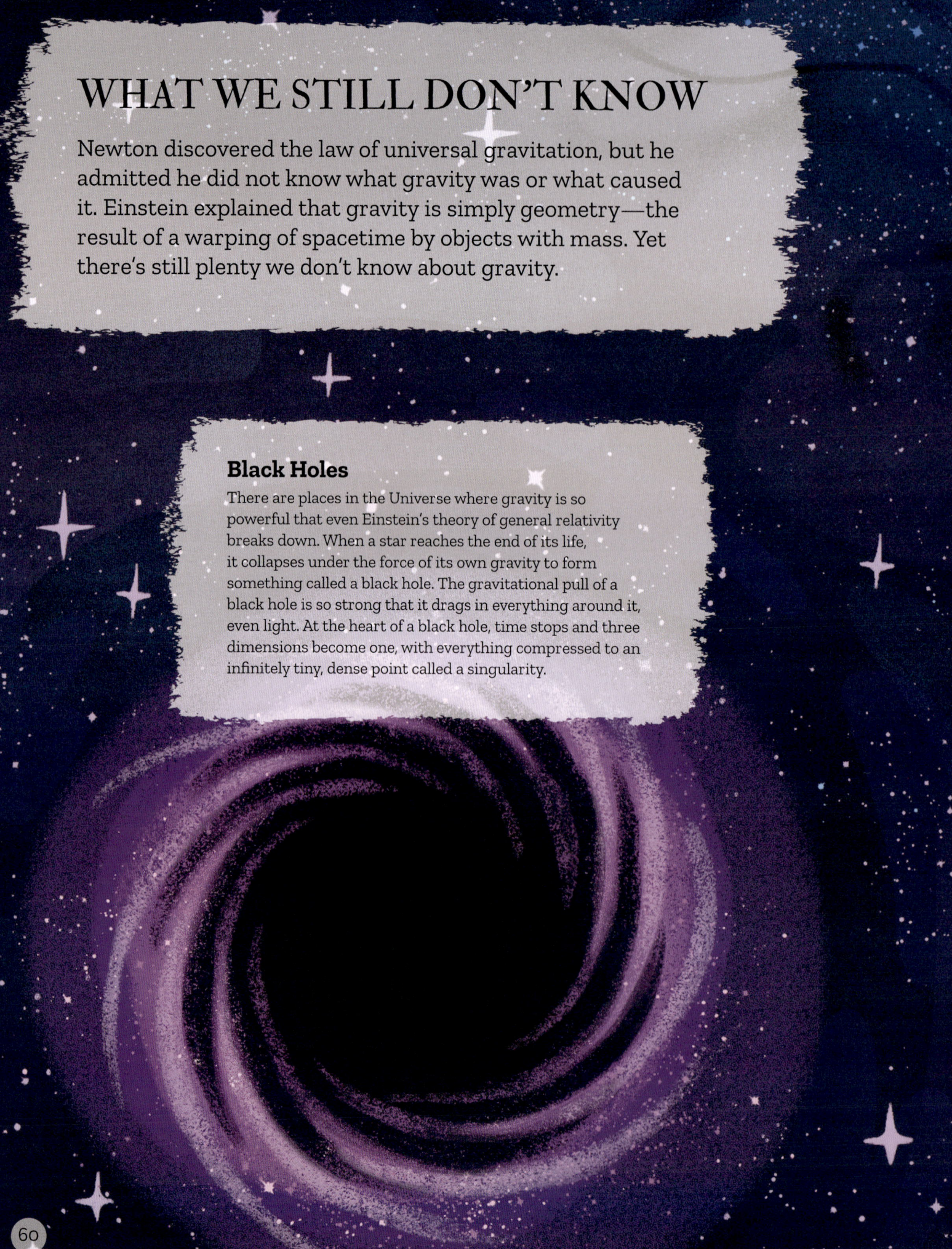

### Black Holes

There are places in the Universe where gravity is so powerful that even Einstein's theory of general relativity breaks down. When a star reaches the end of its life, it collapses under the force of its own gravity to form something called a black hole. The gravitational pull of a black hole is so strong that it drags in everything around it, even light. At the heart of a black hole, time stops and three dimensions become one, with everything compressed to an infinitely tiny, dense point called a singularity.

### Conflicting Theories

For more than 200 years, Newton's laws were dominant. Since the early twentieth century, two new theories have governed physics. General relativity describes the Universe at large scales, where the main force is gravity. Quantum theory describes the Universe at tiny scales, where the three forces are electromagnetism and the strong and weak nuclear forces. These forces are "chunky" (they are carried by particles, or chunks), whereas gravity appears to be "smooth" (it is the curvature of spacetime caused by mass and energy, not particles). Scientists are trying to find a single, unifying theory that describes the Universe at all scales.

THERE ISN'T ENOUGH MATTER IN THE UNIVERSE TO EXPLAIN ALL THE GRAVITY. SOME SCIENTISTS BELIEVE THERE MUST BE AN INVISIBLE SUBSTANCE, CALLED DARK MATTER, THAT'S KEEPING THE GALAXIES FROM FLYING APART.

### The Great Ocean

We are still very far from understanding how gravity, or indeed the Universe, works. Newton was humble enough to recognize this. He said, "I seem to have been only like a boy playing on the seashore, and diverting myself in now and then finding a smoother pebble or a prettier shell than ordinary, whilst the great ocean of truth lay all undiscovered before me."

# GLOSSARY

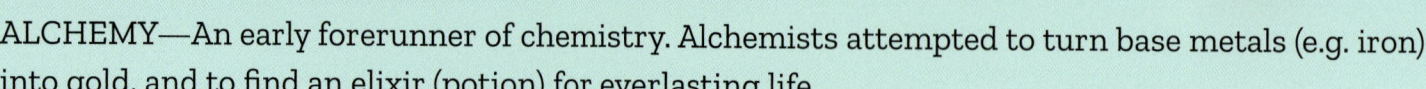

ABSOLUTE—Existing independently and not in relation to other things.

ACCELERATION—A change in speed and/or direction.

ALCHEMY—An early forerunner of chemistry. Alchemists attempted to turn base metals (e.g. iron) into gold, and to find an elixir (potion) for everlasting life.

ASTRONOMER—A scientist who studies space, celestial bodies, and the Universe as a whole.

ATOM—The smallest part of a chemical element.

CALCULUS—A branch of mathematics that can calculate rates of change and motion.

CELESTIAL BODIES—Objects in space, such as stars, planets, moons, and asteroids.

COMET—A celestial object with a core of ice and dust. When near the Sun, it has a tail of gas and dust particles that point away from the Sun.

CURVATURE—The way something is curved.

ELECTROMAGNETISM—The interaction of electricity and magnetism, which gives rise to radiation, including light.

ELECTRON—A subatomic particle with a negative charge. It is found in all atoms.

EQUATION—A statement that the values of two mathematical expressions are equal, shown by the sign "=" (e.g. $F = ma$).

EQUATOR—An imaginary line drawn around Earth's middle, equally distant from the North and South Poles.

ETHER—An invisible substance once thought to fill all of space.

FINITE—Limited in number or size.

FORCE—In physics, this refers to anything that changes the motion of a body or causes a stationary body to move.

FORMULA—A method or procedure for achieving something.

GEOMETRY—The branch of mathematics concerned with the properties and relations of points, lines, surfaces, and solids.

GRAVITY—A force of attraction between objects with mass caused by a warping of spacetime.

INFINITE—Limitless in number or size.

INVERSE—Opposite.

LENS—A piece of glass with curved sides for focusing or scattering light rays, used in telescopes, magnifying glasses, and spectacles.

MAGNIFY—Make something appear larger, for example by viewing it through a telescope.

MASS—The quantity of matter, or physical stuff, that a body contains.

NATURAL PHILOSOPHY—An old term for science.

OPTICS—The science of sight and the way light behaves.

ORBIT—The curved path of an object around a star, planet, or moon.

PARALLEL—Lines running side by side that always have the same distance between them.

PHOTON—A particle of light.

PHYSICS—The branch of science concerned with the nature and properties of matter and energy.

PRISM—A transparent object, usually triangular in shape, that can refract white light into the colors that it is made up from.

QUANTUM MECHANICS—The branch of science that deals with the motion and interaction of subatomic particles.

RATIONAL—Based on reason or logic.

RELATIVITY—The dependence of things like space, time, and gravity on the relative motion of the observer.

SCIENTIFIC METHOD—A procedure for reaching conclusions in science, involving observation, measurement, and experiment.

SOLAR ECLIPSE—An obscuring of the Sun by the Moon that blocks out the Sun's light.

SOLAR SYSTEM—The Sun and the planets and other bodies that circle it.

SQUARE—In mathematics, this is the product (total) of a number multiplied by itself. For example, 9 is the square of 3.

STATIONARY—Motionless.

SUBATOMIC PARTICLE—A particle smaller than or occurring within an atom.

WARP—Bend, curve, or twist.

# INDEX

absolute space and time 54, 55, 57
acceleration 7, 31, 49, 50, 51
al-Biruni 6
alchemy 17, 38–39
Apollo missions 49
apple (and gravity) 4, 20, 21, 22, 26, 28, 29, 30, 32, 41, 56
Aristotle 6, 7, 10
asteroids 29
astronauts 23
astronomy 18, 36, 37, 48, 49, 57
atoms 38, 42, 58, 59
Ayscough, Hannah 8, 9, 10, 12

Bacon, Roger 52
Barrow, Isaac 13, 14
black holes 60
Brahmagupta 6

calculus 13, 18, 32–33
Cambridge University 10, 11, 12, 13, 14
Cassinis, Dominique and Jacques 44, 45
celestial bodies 5
centrifugal force 45
centripetal force 25
Clark, William 9, 10
classical mechanics 42, 58
color 13, 14, 34, 35, 36, 52
comets 48
Copernicus, Nicolaus 10

dark matter 61
deduction 41
Descartes, René 7, 10, 40, 41
double-slit experiment 58
du Châtelet, Émilie 40

Earth 4, 5, 6, 10, 20, 21, 22, 23, 26, 42, 44, 45, 46, 47, 53, 57
Earth's shape 44–45
Eddington, Arthur 57
Einstein, Albert 42, 55, 56, 57, 58, 60
electrons 59
engineering 50–51
Enlightenment 43
ether 7, 40
experiments 7, 12, 13, 22, 25, 39, 52, 53, 55, 58

Flamsteed, John 18
flight 51

force and forces 4, 5, 7, 16, 20, 21, 25, 28, 29, 30, 31, 38, 42, 49, 50, 51, 56
French Revolution 43
friction 30

galaxies 5, 61
Galilei, Galileo 7, 10, 24, 26
general theory of relativity 42, 56, 57, 60, 61
Gravitational Constant 29
gravity 4–7, 11, 13, 16, 20–29, 30, 32, 38, 40, 45, 46, 47, 49, 50, 51, 53, 56, 60, 61

Halley, Edmond 48
Halley's Comet 48
Hooke, Robert 15, 16, 26

induction 41, 53
infinity 32, 33
inverse square law 26–27, 29

James Webb Space Telescope 37

Kepler, Johannes 10, 24, 27

Leibniz, Gottfried 18
law of universal gravitation 20, 29, 41, 42, 46, 48, 49, 50, 51, 56, 60
laws of motion 30–31, 42, 48, 49, 50, 51, 53
lenses 14, 36
light 13, 14, 15, 34–35, 36, 42, 52, 57, 58, 60
light, speed of 42, 55

magnetism 5
mass 4, 7, 16, 26, 28, 29, 31, 45, 51, 56, 57, 60, 61
mathematics 5, 13, 14, 16, 18, 24, 26, 27, 28, 32–33, 50
matter 11, 17, 28, 29, 41, 59
matter waves 59
method of fluxions 13, 14, 33
Moon 4, 5, 6, 20, 21, 22, 23, 26, 28, 41, 46, 47, 48, 49, 56, 57
Mars 49
Mercury 56

Neptune 49
Newtonianism 43, 58
Newton's cradle 42
Newton's life 8–19

Opticks (1704) 14, 59
optics 14
orbits 4, 20, 21, 23, 24, 41, 49, 53, 56

particle theory of light 15, 35, 58
plague 11, 12
planets 4, 5, 13, 16, 24–25, 26, 29, 40, 41, 48, 49, 53
Principia, the (1687) 16, 17, 18, 23, 24, 27, 29, 30, 43, 48, 53, 54
prisms 13, 34, 35, 36
Pytheas 46

quantum theory 42, 58–59, 61

rainbows 34
reflecting telescope 14, 15, 36–37, 48
refracting telescopes 36, 37
relative space and time 55
Richer, Jean 45
rockets 4
Royal Mint 17
Royal Society 15, 18, 37, 44

satellites 22, 23
Schrödinger, Erwin 59
scientific method 52
Smith, Barnabas 8, 9
solar eclipses 48, 57
sound, speed of 53
space 4, 5, 7, 11, 21, 25, 48–49, 54, 55, 57
spaceflight 48, 49
spacetime 56, 57, 60, 61
special theory of relativity 42, 55
speed 24, 30, 32, 42, 50, 53, 55
stars 5, 18, 19, 29, 36, 42, 57, 60
Sun 4, 10, 24, 25, 40, 41, 46, 47, 48, 56, 57
sundial 8

telescopes 14, 36, 37, 48, 49
tides 4, 41, 46–47
time 54, 55, 57, 60
Trinity College Cambridge 10, 11, 14, 53

Universe 6, 16, 21, 40, 42, 53, 54, 58, 60, 61
Uranus 49

Voltaire 40, 45

wave theory of light 15, 58
weight 4, 5, 45
Woolsthorpe Manor 8, 9, 12, 13, 16, 20, 34

Young, Thomas 58